Air Pollution Control Engineering for Environmental Engineers

Fundamentals of Environmental Engineering

Series Editor
Jeff Kuo
California State University, Fullerton, USA

Geotechnical Engineering for Environmental Engineers
Binod Tiwari, Jeff Kuo

Air Pollution Control Engineering for Environmental Engineers
Jeff Kuo

For more information about this series, please visit:
https://www.crcpress.com/Automation-and-Control-Engineering/book-series/CRCAUTCONENG

Air Pollution Control Engineering for Environmental Engineers

Jeff Kuo

CRC Press
Taylor & Francis Group
Boca Raton London New York

CRC Press is an imprint of the
Taylor & Francis Group, an **informa** business

Published in 2018 by CRC Press
Taylor & Francis Group
6000 Broken Sound Parkway NW, Suite 300
Boca Raton, FL 33487-2742

© 2018 by Taylor & Francis Group, LLC
CRC Press is an imprint of Taylor & Francis Group

No claim to original U.S. Government works
Printed in the United States of America on acid-free paper
10 9 8 7 6 5 4 3 2 1

International Standard Book Number-13: FILL IN 978-1-138-03204-0 (CHOOSE:hardback)

Library of Congress Cataloging-in-Publication Data
Catalog record is available from the Library of Congress

Visit the Taylor & Francis Web site at
http://www.taylorandfrancis.com

and the CRC Press Web site at
http://www.crcpress.com

Contents

v

Preface

My experience has given me an in-depth understanding of what an environmental engineering student and/or a junior environmental professional should know about air quality. Whilst teaching at California State University, Fullerton (CSUF) since 1995, I have worked in several consulting companies and in the Sanitation Districts of Los Angeles County. I have also worked on several externally-funded projects related to air quality recently and in the remote past. I have been teaching a course on air pollution control engineering at CSUF for a long while.

Though I believe that it is important for an engineer to obtain professional licensures, this book is not a preparation book for the Fundamental of Engineering (FE) exam and/or Professional Engineering (PE) exam. However, I took these two exams into consideration when I wrote this book. The specifications of the FE exam reflect what NCEES believe an undergraduate environmental engineering major should have learned from his/her school, while those of the PE exam are what a practicing environmental engineer should know about air quality.

The Office of Air and Radiation (OAR) of the U.S. Environmental Protection Agency (EPA) develops national programs, policies, and regulations for controlling air pollution and radiation exposure. It includes four offices: (1) Office of Air Quality Planning and Standards, (2) Office of Atmospheric Programs, (3) Office of Transportation and Air Quality, and (4) Office of Radiation and Indoor Air. Throughout the years these four offices, along with EPA's Office of Research, have published numerous technical articles that are available to the state, tribal and local regulatory agency personnel as well as the general public. In addition, although courses of EPA's Air Pollution Training Institute (APTI) are mostly restricted to state and local environmental agency personnel, the training materials are available to general public and for educational purposes. Consequently, thousands of pages of valuable information are available. These documents cover a broad range of topics and intended for different types of personnel (e.g., engineers, scientists, inspectors, and policy makers). They are also published by different offices and at different times. All this to say, it is not only laborious but impractical to ask students to read all these documents and extract useful information from them.

I always believe that a student should have learned the basics and common sense of engineering as much as possible from their collegiate courses before he/she starts his/her professional engineering career. Unfortunately, I found that on the contrary, young professionals are not prepared. They may be able to do engineering calculations, but they lack sense and understanding on the meanings of the results and could not justify the solutions and values obtained.

Air quality is the engineering field in the U.S. that uses the International Systems of Units (SI) the most, compared to the other engineering fields. However, units of the U.S. Customary System (e.g., feet, pound, and gallon) are more commonly used in other environmental engineering courses. The U.S. Customary System and SI are commonly mixed in the professional fields. Students and junior engineers are often confused by the mix use of the two measurement systems. For instance, I often ask my students and junior engineers a simple question - "Which is a larger force - one pound or one Newton?". The responses are usually unsatisfactory. This book will adopt the SI as the main measurement system but will be coupled with the US customary units when needed. The main purpose is to help the users of this book develop a good sense of the numerical values that they would for example, be able to easily know the typical air flow velocity in a duct.

There are many textbooks available in the market, and most, if not all, are at least 400 pages thick. Although all the information contained in those books are valuable and useful, it would be nearly impossible for a lecturer to cover all the content in one semester. In addition, students may not be able to grasp and retain all the information that the authors intend to convey. This book is shorter than the others, but it still serves as a sufficient introductory book for an air quality and environmental engineering course of 3-semster units. In a succinct format, this practical engineering book provides well-digested information for both engineering students and junior engineering professionals. My hope and objective is for readers to read this book from cover to cover and gain the common sense needed to be well-equipped and successful in their engineering profession.

Chapter 1

Air Pollutants and Their Adverse Effects

Any substance can become a pollutant to the environment, if it is present in an environmental medium (e.g., water, air, or soil) with an undesirable concentration, in an unacceptable state, and/or at an unwanted location. The presence of oxygen molecules is favorable in many situations; however, they would be considered as pollutants in an anaerobic biological reactor. *Air pollution* is the presence of unwanted material in the air. The term "unwanted material" refers to material concentrations present for a sufficient time and under circumstances that interfere significantly with comfort, health, or welfare of persons, or with the full use and enjoyment of property.

This chapter starts with a discussion on characteristics of common air pollutants (section 1.1). Since there are a variety of them, the air pollutants are often grouped into different categories (section 1.2). To effectively control emissions of air pollutants, we need to know common sources of the emissions (section 1.3). The last section of the chapter (section 1.4) provides brief coverage on adverse effects of air pollution.

1.1 Characteristics of Air Pollutants

Air pollutants can be in the form of gas, liquid, or solid. They can be organic, inorganic, or microbial species.

1.1.1 Gas, vapor, particulate matter, soot, smoke, fog, aerosols and smog

Many terms are being used to describe the state of an air pollutant. Air is a mixture of gases. *Gas* is a state of matter that has no fixed shape and volume. *Vapor* refers to a gas phase where the same substance also exists in its liquid and/or solid state under that condition. For example, we say that our ambient air contains nitrogen gas and water vapor. It is because no liquid or solid nitrogen coexists with the nitrogen gas, while water and/or ice is present with water vapor under ambient conditions.

In addition to gases and vapors, air also contains *particulate matter* (PM) which includes suspended solid and liquid, such as dust, soot, smoke, and liquid droplets as well as living matters such as pollen, bacteria, spores, mold, and fungus. *Dust* is a loose term applied to solid particles, predominantly larger than colloids and capable of temporary suspension in air or other gases. Dusts do not diffuse, but settle under the influence of gravity. *Soot* is comprised of particles of amorphous carbon and tars generated from incomplete combustion of hydrocarbons.

An *aerosol* is a mixture of fine PM (liquid or solid) and air, and/or gas. Aerosols can be natural (e.g., fog and geyser steam) or artificial (e.g., haze). *Fog* is a loose term applied to visible aerosols in which the dispersed phase is liquid, mainly formed by condensation. *Smoke* is an aerosol which is a visible mixture of gases and fine particles generated from combustion. The particles are present in sufficient quantity to be observable independently of the presence of other solids. *Smog* is a term derived from smoke and fog, and it was first used around 1950 to describe the combination of smoke and fog in London. *Photochemical smog* is air contamination caused by photochemical reactions among nitrogen oxides and hydrocarbon by the action of sunlight. Ozone is its main component.

1.1.2 Inorganics, organics, or microbial

Air pollutants can be pure inorganic or organic molecules such as nitrogen dioxide (NO_2) and benzene (C_6H_6). They can also be inorganic minerals such as asbestos or living matters such as bacteria. It should be mentioned that an air pollutant can be a combination of two or more pollutants. For example, toxic heavy metals such as lead (Pb) can get adsorbed onto PM and sulfur dioxide (SO_2) can get absorbed into tiny water droplets which may make the PM and the water droplets more toxic and/or unacceptable.

1.1.3 Volatile organic compounds (VOCs)

Volatile organic compounds (VOCs) are of great environmental concerns because of their mobility and toxicity. In general, they are organic compounds having high vapor pressures and can be emitted from solids or liquids under ambient conditions. With regards to air pollution, as specifically defined in 40CFR Part 51.100(s), *volatile organic compounds* are any compound of carbon, excluding carbon monoxide, carbon dioxide, carbonic acid, metallic carbides or carbonates, and ammonium carbonate, which participates in atmospheric photochemical reactions. Many organic compounds that are volatile, but they are not considered as VOCs by the U.S. Environmental Protection Agency (referred to as EPA in this book) in air pollution management because of having negligible photochemical reactivity. In addition to chlorofluorocarbons (CFCs), hydrofluorocarbons (HFCs), and hydrochlorofluorocarbons (HCFCs), the other notable exceptions include acetone, dimethyl carbonate, ethane, methane, methyl acetate, 1,1,1-trichloroethane (1,1,1-TCA), methyl formate, methylene chloride, siloxanes, and tetrachloroethylene. Many of these exceptions (e.g., 1,1,1-TCA) are VOCs of concern in water or wastewater treatment; however, they not considered as VOCs in air regulations.

1.1.4 Primary and secondary air pollutants

For drinking water, the EPA imposes primary standards to protect public health and instills secondary standards to regulate substances that may cause negative aesthetic effect. For the ambient air, the EPA also imposes primary and secondary air quality standards. However, with regards to types of air pollutants, the "primary" and "secondary" have different meanings.

A *primary air pollutant* is an air pollutant emitted directly from a source, such as an industrial stack or a vehicle exhaust pipe; examples include carbon monoxide (CO) and nitric oxide (NO). In the atmosphere it has the same physical and chemical forms as when it was emitted. A *secondary air pollutant* is not directly emitted from a source, but formed from chemical or photochemical reactions among the primary air pollutants in the atmosphere. Ozone (O_3) is a good example of secondary pollutants; it is formed from photochemical reactions among sunlight, nitrogen dioxide (NO_2), and hydrocarbons (HCs). Acid rain is another example because it is formed from the reactions between water droplets and nitrogen oxides (NO_x) and/or sulfur oxides (SO_x). Some air pollutants can be both primary and secondary. NO_2 is mainly a secondary air pollutant formed from oxidation of NO in the ambient air, but it can also be a primary air pollutant emitted directly from combustion processes. VOCs can be primary or secondary air pollutants. Larger PMs in the ambient (e.g., dust, volcano ash) are typically primary air pollutants, while finer PMs are typically secondary air pollutants that are generated from physical (e.g., condensation), chemical or photochemical reactions in the atmosphere.

1.2 Groups of Air Pollutants

The EPA categorizes the air pollutants into two groups: outdoor and indoor. The outdoor pollutants are further categorized into (a) criteria air pollutants, (b) hazardous air pollutants (HAPs), and (c) greenhouse gases (GHGs).

1.2.1 Criteria air pollutants

The main federal law of the United States for controlling air quality, the Clean Air Act (CAA), requires the EPA to set National Ambient Air Quality Standards (NAAQS) for six *criteria pollutants* in outdoor air to protect human health, properties, ecosystems, and the environment. They are (1) CO, (2) NO_2, (3) SO_2, (4) ground-level O_3, (5) PM, and (6) Pb. They are present in the ambient air all over the U.S. The term "criteria pollutants" derives from the requirement that the EPA must describe the characteristics and potential health and welfare effects of these pollutants. These standards are set or revised on the basis of these criteria.

1.2.2 Hazardous air pollutants

Hazardous air pollutants (HAPs) are air pollutants which have been listed pursuant to Section 112(b) of the Clean Air Act Amendments (CAAA) of 1990. They are categorized under National Emission Standards for Hazardous Air Pollutants (NESHAPs) or Hazardous Air Pollutant (HAPs) regulations. They are also known as toxic air pollutants or air toxics. *Air toxics* are known or suspected to cause cancer or other serious health effects, such as reproductive and birth as well as adverse environmental effects. The original list included 189 pollutants; since 1990 the EPA has modified the list through rulemaking to 187 HAPs. The majority of the 187 HAPs are volatile organic compounds. In addition to the commonly-mentioned toxic organics (e.g., acetaldehyde, benzene, chlorine, chlorobenzene, chloroform, 1,4-dioxane, formaldehyde, methylene chloride, phenol, styrene, tetrachloroethylene, toluene, and vinyl chloride), pesticides (e.g., chlordane and lindane), heavy metals (e.g., mercury, cadmium, lead and chromium compounds), acids (e.g., hydrochloric acid and hydrofluoric acid), cyanide compounds, and asbestos, the list also includes compounds that are often considered not that toxic such as methanol and phosphorus. The list can be found at https://www.epa.gov/haps/initial-list-hazardous-air-pollutants-modifications. Out of six criteria pollutants, lead is both a criteria pollutant and an air toxic.

1.2.3 Greenhouse gases

Greenhouse gases (GHGs) are those gases can trap heat in the atmosphere. Elevated concentrations of GHGs due to human activities are considered as the main contributor to global warming and climate change. The GHGs of concern are carbon dioxide (CO_2), methane (CH_4), nitrous oxide (N_2O) and fluorinated gases.

1.2.4 Indoor air pollutants

Indoor air pollutants are those within and around buildings and structures that will affect the health and comfort of the building occupants.

1.3 Sources of Air Pollutants

Emissions are the total substances discharged. *Air emission source*s include anything that releases pollutants into air. There are many sources of air emissions and there are many terms used to describe the emission sources. For example, >75% of emissions of criteria pollutants, except ozone, originates from combustion. Combustion, a process, here is the emission source. Some commonly-used ones are described in this section.

1.3.1 Natural, biogenic, and anthropogenic sources
Examples of natural sources of air pollutants include wildfires, wind-blown dust, volcanoes, and vegetation. *Biogenic emissions* are emissions from natural sources, such as plants and trees. These sources emit HAPs (e.g., formaldehyde and acetaldehyde) as well as large quantities of non-HAP VOCs. Examples of anthropogenic sources include combustion, industrial processes, agriculture, and transportation. *Anthropogenic emissions* are resulted from human activities and they are of particular interests because they can be reduced through regulatory and voluntary actions, leading to air quality improvements.

1.3.2 Stationary vs. mobile sources
Air pollutant emissions sources can also be grouped into two categories: stationary sources and mobile sources. A *stationary source* is defined as any building, structure, facility, or installation which emits or may emit any air pollutant. *Mobile sources* include on-road vehicles and non-road vehicles and equipment.

1.3.3 Point, area, line and volume sources
Stationary sources can be further divided into two main subcategories: point and area sources. *Point sources* consist of a single emission source with an identified location point at a facility. Facilities could have multiple point sources located onsite. Point sources are usually associated with manufacturing and industrial processes, such as boilers, spray booths or degreasers. *Area sources* are small emission sources that are widely distributed, but may have substantial cumulative emissions; examples include residential water heaters, small engines, and consumer products, such as barbecue lighter fluid and hair spray.

A *line source* is a source of emissions emanates from a linear (one-dimensional) geometry. A highway can be a line source of air pollution. A *volume source* is a 3-dimensional emission sources. Dust emissions from wind erosion of coal piles and fugitive VOCs emissions from equipment components (e.g., flanges and valves) within an oil refinery are examples of volume sources (see Section 1.3.5 below for definition of fugitive emissions)

1.3.4 Major and area sources
With regards to air toxics, *area sources* are specifically defined as facilities having toxic air emissions below the *major source* threshold, which is 10 tons/year of a single toxic air pollutant or 25 tons/year of multiple toxic air pollutant, defined by the CAA.

1.3.5 Fugitive emissions
Fugitive emissions are emissions that escape from industrial processes and equipment.

1.4 Adverse Effects of Air Pollution

Air pollution has adverse impacts on human health, aesthetics (e.g., visibility and smells), properties, ecosystems, and our global environment. A *receptor* of air pollution is the entity which is exposed to air pollutants. It can be any human, organ of human body or environmental systems that can be affected by air pollution. Human receptors include skin, eyes, nose and the respiratory system. *Sensitive receptors* are people (e.g., children, the elderly people, pregnant woman, and patients) who have an increased sensitivity to air pollution or environmental contaminants. *Sensitive receptor locations* include schools, parks and playgrounds, day care centers, nursing homes, and hospitals. This section briefly describes adverse effects of air pollutants; more details can be found at https://www.epa.gov/environmental-topics/air-topic.

1.4.1 Human health
From inhalation, ingestion, and dermal contact, air pollutants can enter the human body. They may remain in the lung or move into the blood stream that goes through all parts of the body. They may go through chemical changes in the body, especially in the liver, before being exhaled or excreted. The effects of air pollution on human health vary from one person to another. However, the elderly, infants, pregnant women and patients with chronic lung or heart diseases are considered as the most sensitive receptors to the impacts of air pollution. The adverse effects can be acute or chronic.

Lead (Pb). Once inside the human body, Pb is distributed in the blood throughout the body and accumulated in bones. Adverse impacts most commonly encountered are neurological effects in children and cardiovascular effects in adults. Infants and young children are especially sensitive to lead exposures, even at low levels, which may contribute to their behavioral problems, learning deficits and lower IQs.

Carbon monoxide (CO). A high CO concentration in inhaled air will reduce the amount of oxygen that can be transported in the blood stream to the vital organs (i.e., heart and brain). At very high levels, CO can cause dizziness, confusion, unconsciousness and even death. It poses extraordinary threats to people with some types of heart disease.

Ground-level O_3. Ozone can cause muscles in the airways in the respiratory system to constrict, trapping air in the alveoli that leads to wheezing and shortness of breath. Long-term exposure to ozone may cause aggravation of asthma as well as permanent lung disease. Children, older adults, outdoor workers as well as people with reduced intake of certain nutrients (e.g., vitamins C and E), with certain genetic characteristics, or with asthma, are at great risk from ozone exposures.

Nitrogen dioxide (NO_2). Breathing air with high NO_2 concentrations can irritate airways in the respiratory system. Short-term exposures to elevated levels of NO_x can aggravate respiratory diseases, particularly asthma. Longer-term exposures may contribute to development of asthma and potentially increase susceptibility to respiratory effects. Children, the elderly, and people with asthma are generally at greater risk from NO_2 exposure. NO_2 can react with other chemicals in the atmosphere to form particulate matter and ozone, the other two criteria pollutants.

Sulfur dioxide (SO_2). Short-term exposures to elevated levels of SO_2 can harm the respiratory system and make breathing difficult. Children, the elderly, and people with asthma are generally at greater risk from SO_2 exposure. SO_2 can react with other chemicals in the atmosphere to form particulate matter, another criteria pollutant.

Particulate matter (PM). PM can accumulate in the respiratory system to cause numerous adverse health effects, especially related to the lung and heart. Particles larger than 10 micrometers (μm, or micron) can be retained in the upper part of the respiratory system where they are easily expelled by physiological processes, while the smaller ones can enter deeper into the system. The liquid particles and soluble solid particles may be absorbed into body tissues and pose additional health risks, including premature death for people with heart or lung disease, nonfatal heart attacks, irregular heartbeat, aggravated asthma, decreased lung function, and increased respiratory problems. Children, the elderly, and people with heart or lung diseases are generally at greater risk from PM exposure.

Hazardous air pollutants (HAPs). People exposed to HAPs at sufficient concentrations may have an increased chance of getting cancer or experiencing other series health effects, including damage to the immune systems, as well as neurological, reproductive, developmental, respiratory and other health problems.

1.4.2 Ecosystems
Lead. Lead can enter soils and sediments from air through deposition and it is persistent in the environment. Elevated levels of lead in the environment can result in decreased growth and reproductive rates in animals and plants as well as neurological effects in vertebrates.

Ground-level O_3. Ozone can have adverse effects on sensitive vegetation by reducing photosynthesis, slowing plants' growth and increasing their risks of disease, insect damages, effects of other pollutants and harm from severe weather. These impacts on sensitive plants can then have negative impacts on ecosystems, including loss of species diversity as well as changes to habitat quality, water and nutrient cycles, and the specific assortment of plants present in a forest.

Nitrogen dioxide (NO_2). NO_2 can react with other chemicals in the atmosphere to form acid rain which harms sensitive ecosystems such as lakes and forests. The nitrate particles originated from NO_2 make air hazy and reduce visibility. NO_2 in the atmosphere can also contribute to nutrient pollution in coastal waters.

Sulfur dioxide (SO_2). At high concentrations, gaseous SO_2 can harm vegetation by damaging foliage and decreasing growth. Sulfur dioxide by reacting with other chemicals in the atmosphere to form acid rain which harms sensitive ecosystems such as lakes and forests. The particles originated from SO_2 make air hazy and reduce visibility.

Particulate matter (PM). Fine particles ($PM_{2.5}$) are the main cause of haze and reduced visibility in parts of the U.S. PM can be carried long distances and then settle on ground or surface water. The adverse effects of this deposition could (i) make surface water acidic and change its nutrient balance, (ii) deplete nutrients in soil, (iii) harm sensitive forests and farm crops, (iv) affect the diversity of ecosystems, and (v) contribute to acid rain effects.

Hazardous air pollutants (HAPs). Like humans, animals may experience health problems if taking in sufficient doses of air toxics. In addition, some toxic air pollutants (e.g., mercury) can deposit into soil or surface water, where they can be extracted by vegetation and ingested by animals and eventually get into the food chain.

1.4.3 Properties

Deposition of particles as well as acid rain can stain and damage stone and other materials, including faces of buildings and culturally important objects such as statues and monuments.

1.4.4 Global environment

Acid rain can cause problems beyond the regional and national scale. Air pollution can be on a global scale such as ozone layer depletion, global warming, and climate change.

1.4.5 Useful data sources

EPA's Report on the Environment (ROE) shows how conditions of the U.S. environment and human health are changing over time (https://cfpub.epa.gov/roe/). The ROE presents indicators of national trends in five theme areas: air, water, land, human exposure and health, and ecological conditions. The ROE air indicators address three fundamental questions: (1) what are the trends in outdoor air quality and their effects on human health and the environment? (2) what are the trends in greenhouse gas emissions and concentrations and their impacts on human health and the environment? (3) what are the trends in indoor air quality and their effects on the human health? Most of the trends are presented graphically and easy to comprehend.

National-Scale Air Toxics Assessment (NATA) is EPA's ongoing comprehensive evaluation of air toxics in the U.S. NATA contains recent emissions data and it uses models to make broad estimates of health risks over geographic areas of the country (https://www.epa.gov/national-air-toxics-assessment).

Exercises

1. With regards to anthropogenic mercury emissions in the United States,
   (a) What are the major source categories?
   (b) Which sector is the largest emission source?

(c) What is the total emission of the most recent year available on the website of EPA's Report on the Environment?

(d) How much reduction (in percentage) in total mercury emissions from 1990 to the most recent year?

2. (a) What are the median and average ambient 8-hr ozone concentrations in the United States in the most recent year available on the website of EPA's Report on the Environment?

   (b) How much reduction (in percentage) in median concentrations from 1978 to the most recent year?

3. (a) What are "current asthma prevalence" and "asthma attack prevalence"?

   (b) What is the age-adjusted asthma attack prevalence in U.S. adults (18+ years) in the most recent year available on the website of EPA's Report on the Environment?

   (c) How much reduction in asthma attack prevalence (in percentage) from 2002 to the most recent year?

Chapter 2

Properties of Air and Compounds of Concern

Having a good understanding on properties of air and compounds of concern (COCs) and interactions among them is vital to establish and maintain an effective air quality program. The atmospheric conditions vary by locations, by time of the day, and by season, to name a few. Operating conditions inside a control device are often different from the ambient conditions. Understanding how environmental or operating conditions affect properties of air and COCs and interactions among them are important in engineering practices.

This chapter starts with a coverage on the Earth's atmosphere (Section 2.1), followed by factors affecting properties of air (Section 2.2), and then basic properties of air, including molecular weight, density, viscosity, specific heat capacity, and Reynolds number (Section 2.3). Section 2.4 presents some important characteristics of COCs, (e.g., vapor pressure, humidity, and heating value) that are relevant to air pollution and control. Section 2.5 discusses the interactions among gas, liquid, and solid which are the phases of air pollutants and their hosting media in the gas stream (and in the atmosphere) as well as in control/removal devices.

2.1 Our Atmosphere

Atmosphere of the Earth is a mixture of many different gases held by the gravitational force. Major components of dry air include nitrogen (78.08% by volume or by mole), oxygen (20.95%), and argon (0.93%). Carbon dioxide (CO_2), the major greenhouse gas (GHG), accounts for most of the balance (0.04%) with other gases in trace amounts. Water vapor concentrations vary at locations. Our atmosphere also contains particulate matter (PM).

Earth's atmosphere is about 480 kilometers (300 miles) in thickness. Atmospheric pressure decreases as the altitude increases due to the decreasing mass of air above. The atmosphere can be divided into four layers based on its temperature; and they are troposphere, stratosphere, mesosphere and thermosphere. Beyond the atmosphere, there is the *exosphere*. Figure 2.1 illustrates the temperature profile of the atmosphere.

Troposphere, the lowest part of the atmosphere, starts from the Earth's surface to 8 to 14.5 km high (5 to 9 miles). It is the part of the atmosphere that we live in. Its thickness is less than 3% of the entire atmosphere; however, it contains about 75% of all the atmospheric mass and almost all of the water vapor. In the troposphere, its temperature decreases with altitude. The bottom part of the troposphere is called the *planetary boundary layer* in which the air motion is affected by the Earth's surface. The air in the troposphere will move horizontally and vertically and redistributes heat, moisture and other constituents, including air pollutants along with its movement. The upper limit of the troposphere is called *tropopause* in which the temperature is relatively constant. Clouds and weather generally occur below the tropopause in the troposphere. The airlines often fly around the tropopause.

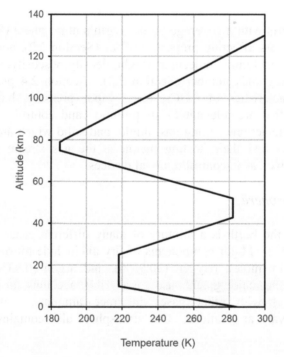

Figure 2.1 - Temperature profile of the atmosphere

The *stratosphere* starts above the troposphere and ends about 50 km (31 mi) and it contains much of the ozone in the atmosphere. With regards to air pollution, the troposphere is the layer where we have most of our daily activities as well as almost all the weather in; while the ozone layer, which absorbs harmful ultraviolet (UV) radiation from the Sun, is within the stratosphere. The temperature in this layer increases with altitude because absorption of solar UV radiation.

The layer above the stratosphere is the *mesosphere*. The temperature decreases with height, reaching a minimum temperature of about 185K at the *mesopause* from the Earth's surface. The *thermosphere* is the top layer of the atmosphere. Its temperature increases with altitude because of absorption of solar UV and X-ray radiation. The region of the atmosphere above about 80 km (50 mi) is called *ionosphere*, since the energetic solar radiation would knock electrons off molecules and atoms in this region.

2.2 Factors Affecting Properties of Air

This section talks about temperature and pressure (two of the important parameters affecting air properties), the ideal gas law, and the standard conditions.

2.2.1 Temperature

In our daily activities, we express temperatures (T) in degree Fahrenheit (^{o}F) or degree Celsius (^{o}C). A temperature can be converted from degree Celsius to degree Fahrenheit (or vice versa) by:

$$T \ (in \ ^oF) = 1.8 \times T \ (in \ ^oC) + 32 \qquad (2.1)$$

However, when temperature is part of an equation describing a physical or chemical relationship (e.g., the Ideal Gas Law), the temperature should be the absolute temperature. The *absolute temperature* is a temperature scale that starts with zero at absolute zero temperature (i.e., the temperature at which kinetic energy is negligible). For the International System (SI) of units, degree Kelvin (K) is used to express an absolute temperature. The following equation converts a temperature in degree Celsius to Kelvin:

$$T \ (in \ K) = T \ (in \ ^oC) + 273.15 \approx T \ (in \ ^oC) + 273 \qquad (2.2)$$

For the US customary units, absolute temperatures are expressed in degree Rankine (^{o}R). Eq. 2.3 converts a temperature from degree Fahrenheit to degree Rankine:

$$T \ (in \ ^oR) = T \ (in \ ^oF) + 459.67 \approx T \ (in \ ^oF) + 460 \qquad (2.3)$$

Therefore, water boils at 100 ^{o}C, 212 ^{o}F, 373K, or 672 ^{o}R, when pressure (P) = 1 atm. A change of temperature in one ^{o}C or one K is equal to a change of

13

1.8 °F or 1.8 °R. The conversion for a temperature in degrees Rankine and Kelvin is simple and straight-forward:

$$T \ (in \ ^{o}R) \ = \ 1.8 \times T \ (in \ K) \tag{2.4}$$

2.2.2 Pressure

A stationary object may be subjected to tensile and/or compressive forces. When it is in motion, it will also be subjected to shear/friction force. *Pressure/stress* is defined as force per unit area. Pressure is also an important parameter affecting properties of air and gas/vapor. *Pressure drop* (*head-loss*) is a matter of concern for air movement in treatment units (e.g., absorber) and transport systems (e.g., ductwork and blower). A larger head-loss means that we need to provide more energy to overcome it and it will incur a larger cost.

Standard barometric or atmospheric pressure is the average atmospheric pressure at sea level, 45° north latitude at 35 °F (~275 K) and that is equal to one atmosphere (atm). The barometric pressure at a location would depend on its altitude, latitude and weather conditions. The most commonly-used SI and U.S. Customary units for pressure are Newton/m² (N/m² or Pascal (Pa)) and lb/in² (psi), respectively. However, many different units are being used and an engineer often needs to convert a pressure value from one unit to another. Common units of pressure and their relationships are given below. It would be a good idea to memorize them or have them readily available.

1 atmosphere (atm)

$= 1.01325 \times 10^{5} \ N/m^{2} = 1.01325 \times 10^{5} \ Pa = 101.325 \ kN/m^{2} \approx 101.3 \ kN/m^{2}$

$= 1.013 \ bar = 1,013 \ milli-bar \ (mbar)$

$= 14.696 \ lb/in^{2} \ (psi) \approx 14.7 \ psi$

$= 760 \ millimeters \ of \ mercury \ (mm-Hg) = 29.92 \ in-Hg = 760 \ Torr$

$= 33.9 \ ft-H_{2}O = 407 \ in-H_{2}O = 10.33 \ m-H_{2}O$

A reading of a pressure gauge tells the difference in pressure between the system and its surrounding. When the system pressure is higher than its surrounding, the *gauge pressure* is positive. The gauge pressure will have a negative value when the system is under vacuum. The *absolute pressure* (P_{abs}) is the sum of the barometric or atmospheric pressure (P_b) and the gauge pressure (P_g):

$$P_{abs} \ = \ P_b + Pg \tag{2.5}$$

14

2.2.3 Properties of ideal gases

For ideal gases, there are no inter-molecular forces between gas molecules and these molecules occupy no volume. Actually, there is no such a thing as an ideal gas in reality. However, behaviors of real gases deviate insignificantly from those of an ideal gas under ambient P and T.

Boyle's Law states that the volume (V) of an ideal gas is inversely proportional to system P, when the system T is held constant. *Charles' Law* states that P is proportional to T when the V is held constant. Combining these two laws yields the Ideal Gas Law:

$$PV = \left(\frac{m}{MW}\right) RT = nRT \qquad (2.6)$$

where m = mass, MW = molecular weight, and n = number of moles of the gas, and R = the universal gas constant. Below are the values of R in some commonly-used units:

$$R = 0.082 \; \frac{L \cdot atm}{g-mole \cdot K} = 8.314 \frac{m^3 \cdot P_a}{g-mole \cdot K} = 10.731 \frac{ft^3 \cdot psi}{lb-mole \cdot \,^oR}$$

Example 2.1 Molar volume of an ideal gas

Find the volume of 1 g-mole and 1 lb-mole of an ideal gas at T = 0 °C and P = 1 atm.

Solution:

(a) For 1 g-mole of an ideal gas,

$$V = \frac{nRT}{P} = \left[(1 \; g-mole)\left(0.082 \; \frac{L \cdot atm}{g-mole \cdot K}\right)(273 \; K)\right] \div 1 \; atm = 22.4 \; L$$

(b) For 1-lb mole of an ideal gas,

$$V = \left[(1 \; lb-mole)\left(10.731 \frac{ft^3 \cdot psi}{lb-mole \cdot \,^oR}\right)(492 \;\,^oR)\right] \div 14.7 \; psi = 359 \; ft^3$$

Discussion: Using similar approaches, one can readily find that volumes of one g-mole ideal gas at T = 20 and 25 °C (P = 1 atm) are 24.05 and 24.46 liters, respectively; while those of 1 lb-mole ideal gas would be 385 and 392 ft³ at T = 68 and 77 °F, respectively. It may not be a bad idea to have these values readily available or memorized.

Standard temperature and pressure (STP) is a set of conditions for experimental measurements so that comparisons can be made between data taken under different conditions. In chemistry, the STP is defined by International Union of Pure and Applied Chemistry (IUPAC) as T = 273.15 K and P = 100 kPa. However, the standard conditions used in practice are often different. For example, the EPA generally has 760 mm-Hg (or 1 atm) as the standard P, but the standard temperature for stack sampling is 20 °C (68 °F) and that for ambient air sampling and combustion analysis is 25 °C (77 °F). The South Coast Air Quality Management District (SCAQMD) in southern California has 60 °F or 68 °F as the standard temperatures in their regulations. It is always a good practice to state the conditions of STP in data reporting. In this book P = 1 atm and T = 20 °C are used as the standard conditions, unless otherwise specified.

2.3 Basic Properties of Air

This section discusses some basic air properties including molecular weight, density, viscosity, specific heat capacity, and Reynolds number.

2.3.1 Molecular weight

Since air is a mixture of many gases, the (average) MW of air can be estimated from its composition. For most air pollution calculations, it is often satisfactory to consider that the composition of dry air is 21% O_2 and 79% N_2 by volume (also by mole). Thus,

$MW_{dry\ air} \approx [(32)(21\%) + (28)(79\%)] \approx 29$ g/g-mole (lb/lb-mole)

Molar ratio of $N_2{:}O_2$ in air $\approx (79\%) \div (21\%) \approx 3.76$

Mass ratio of $N_2{:}O_2$ in air $\approx [(28)(79\%)] \div [(32)(21\%)] \approx 3.29$

For wet air, the apparent MW can be found by including the volume (or molar) fraction of water (f_{water}). Thus,

$$MW_{wet\ air} = (1 - f_{water})(MW_{dry\ air}) + (f_{water})(MW_{water}) \quad (2.7)$$

2.3.2 Density

Density (ρ) of a material is the ratio of its mass (m) and the volume (V) it occupies. For an ideal gas:

$$\rho = \left(\frac{m}{V}\right) = \frac{P \times MW}{R \times T} \quad\quad (2.8)$$

16

Example 2.2 Density of air

Determine the density of air (a) under the standard conditions (T = 20 °C and P = 1 atm), (b) T = 200 °C and P = 1 atm, and (c) T = 200 °C and P = 2 atm.

Solution:

(a) Assuming MW of air = 29, then

$$\rho = \frac{(1\ atm)(29\frac{g}{g-mole})}{(0.082\frac{L\cdot atm}{g-mole\cdot K})(293\ K)} = 1.21\ g/L$$

or

$$\rho = \frac{(14.7\ psi)(29\frac{lb}{lb-mole})}{(10.731\frac{ft^3\cdot psi}{lb-mole\cdot\ °R})(528\ °R)} = 0.075\ lb/ft^3$$

(b) At T = 200 °C and P = 1 atm:

$$\rho = \frac{(1)(29)}{(0.082)(473\ K)} = 0.75\ g/L$$

(c) At T = 200 °C and P = 2 atm:

$$\rho = \frac{(2)(29)}{(0.082)(473\ K)} = 1.50\ g/L$$

Discussion:

1. Since the molar volume of an ideal gas is 24.05 L at T = 20 °C and P = 1 atm, its density can be readily calculated as (29)/(24.05) = 1.21 g/L.

2. Under these conditions, the density of air (1.21 g/L) is ~800 times smaller than that of water, which is 1 kg/L.

3. Since the volume of an ideal gas increases with temperature, its density will decrease with temperature. The ratio of densities at two different temperatures can be directly obtained from the inverse of the ratio of two absolute temperatures.

4. The density of an ideal gas is directly proportional to the system pressure.

2.3.3 Viscosity

Viscosity of a fluid (i.e., air or liquid) is a measure of its resistance to flow. This resistance is due to inter-molecular friction between layers of fluid when they slide against each other. The shearing stress (τ_{shear}) between layers of a

17

fluid is related to the velocity gradient *(du/dy)* between the liquid layers (see equation below). [Note: A *gradient* is the change in value of a quantity (e.g., concentration) related to the change in a given variable (usually distance). It can be considered as a driving force to cause a movement from one location to another. In environmental engineering applications, temperature, pressure, concentration, and hydraulic gradients are commonly encountered.]

$$\tau_{shear} = \mu\left(\frac{du}{dy}\right) \qquad (2.9)$$

where μ is the *dynamic/absolute viscosity*, or *viscosity*. Common units of viscosity in SI are N-s/m^2 or Poise; while that of the U.S. customary system is lb$_f$-s/ft^2. Conversions among commonly-used units are:

- 1 N-s/m^2 = 1 kg/(m-s) = 10 Poise = 0.0209 lb$_f$-s/ft^2 = 0.0209 slug/(ft-s)

- 1 Poise = 100 centi-Poise (cp) = 0.0672 lb$_m$/ft-s

In engineering applications, *kinematic viscosity* (*v*) is also used, which is the ratio of the absolute viscosity (μ) and the mass density (ρ) of the fluid (i.e., $v = \mu/\rho$). The common units of SI for kinematic viscosity are m^2/s, Stoke (St), or centi-Stoke (cSt); while that of the U.S. customary system is ft^2/s.

- 1 St = 100 cSt = 1 cm^2/s = 10^{-4} m^2/s = 1.08 × 10^{-3} ft^2/s

Table 2.1 - Viscosities of air and water (T = 293K & P = 1atm)

	Viscosity (N-s/m^2)	Kinematic viscosity (m^2/s)
Air	1.81 × 10^{-5}	1.50 × 10^{-5}
Water	1.01 × 10^{-3}	1.01 × 10^{-6}

Because gas molecules are far apart for inter-molecular cohesion to take place, gas is much less viscous than water. The absolute/dynamic viscosity values of water and air at 293 K are 1.01 × 10^{-3} and 1.81 × 10^{-5} N-s/m^2, respectively (Table 2.1). Thus, water is ~55 times more viscous than air under ambient conditions. However, it should be noted that the kinematic viscosity of air at STP is larger than that of water (1.50 × 10^{-5} vs. 1.01 × 10^{-6} m^2/s), mainly because the density of air is ~800 times smaller.

Viscosity of a fluid is relatively independent of pressure, but depends significantly on temperature. Liquid viscosity decreases when the temperature increases; the gas has the opposite trend because the increasing temperature would cause more collisions among the gas molecules. The viscosity of air at any temperature can be estimated by using the viscosity value at a reference absolute temperature (μ_{ref}) and the equation below (EPA, 2012). The viscosity value of 1.81×10^{-5} N-s/m^2 in Table 2.1 can be used as the reference value at T = 293K.

$$\frac{\mu}{\mu_{ref}} = \left(\frac{T}{T_{ref}}\right)^{0.768} \qquad (2.10)$$

Example 2.3 Viscosity of air

Estimate the viscosity of air at T = 200 °C and P = 1 atm.

Solution:

$$\frac{\mu}{\mu_{ref}} = \left(\frac{T}{T_{ref}}\right)^{0.768} \rightarrow \frac{\mu}{1.81 \times 10^{-5}} = \left(\frac{473}{293}\right)^{0.768} \qquad *2.61467 \times 10^{-5}?$$

$$\mu_{473K} = 4.73 \times 10^{-5} \text{ kg/(m-s)} = 9.89 \times 10^{-7} \text{ lb}_f\text{-s/ft}^2$$

Discussion: As expected, viscosity of air increases with temperature.

2.3.4 Specific heat capacity

Specific heat capacity (or *specific heat*) of a gas is the amount of heat needed to raise the temperature of a unit-mass by one degree. Common units of specific heat are J/kg-K, Btu/lb-°F, or calories/g-°C. The conversions among these units are:

- 1 Btu/lb-°F = 1 kcal/kg-°C = 4.184 kJ/kg-K
- 1 kJ/kg-K = 0.239 Btu/lb-°F

Heat can be added to a gas mass while the volume or pressure of the system is kept constant, so that there are *specific heat at constant volume* (C_v) and *specific heat at constant pressure* (C_p). For an ideal gas,

$$C_p = C_v + R \qquad (2.11)$$

where R is the universal gas constant. Commonly-used energy units are Joule (J), calorie, and British thermal unit (BTU). Joule is the unit of work/energy in the SI. It is the work done on an object when a force of one Newton acts on that object in the direction of its motion for a distance of one meter (= 1 N-m). One calorie is the energy needed to raise the temperature of 1 gram of water by 1 °C, while one Btu is the energy needed to raise the temperature of 1 pound of water by 1 °F at the temperature when water has its greatest density (~39 °F or ~ 4 °C). One Joule is about one quarter of one calorie and is about one thousandth of one Btu. Conversions among these common units are:

- 1 J = 1 N-m = 9.45×10^{-4} Btu
- 1 Btu = 1,055 J
- 1 J = 0.239 calorie (1 calorie = 4.184 J)

From the definitions of calorie and Btu, we should be able to tell the heat capacity of water under room temperatures is equal to or about 1 Btu/lb-°F, 1 cal/g-°C, 1 kcal/kg-°C, or 4.184 kJ/kg-K. The specific heat of a gas stream depends on its composition and temperature. Table 2.2 tabulates the specific heat of air, CO_2, and water vapor at several temperatures. For a temperature not listed in this table, an estimate can be made by interpolating between two known values. As shown, the heat capacity increases with temperature. The heat capacity values of water vapor are approximately twice of those of air and CO_2. Consequently, the heat capacity of a gas stream contains a large amount of water vapor that would be significantly higher than that of a drier air. In addition, it would take more heat to raise the temperature of a gas stream when it is wet.

Most of energy-related calculations for gas combustion deal with flue gases at a temperature ≤ 1,500 °F (815 °C). For those cases, heat capacity of dry gas is about 0.25 Btu/lb-°F (1.05 kJ/kg-K) and that of water vapor is about 0.50 Btu/lb-°F (2.1 kJ/kg-K). These values can be used for most calculations without introducing significant errors (EPA, 2012).

Table 2.2 - Specific heat capacity of air, CO_2, and water vapor, in Btu/lb-°F or kcal/kg-°C (EPA, 2012)

Temperature		Specific Heat Capacity		
(°F)	(°C)	Air	CO_2	Water Vapor
68	20	0.242	0.200	0.445
212	100	0.244	0.218	0.452
500	260	0.249	0.245	0.470
1,100	593	0.260	0.285	0.526
2,200	1,204	0.278	0.315	0.622
3,000	1,649	0.297	0.325	0.673
3,800	2,093	0.303	0.330	0.709

Example 2.4 Specific heat capacity of a gas stream

Estimate the specific heat capacity of an exhaust gas stream with 15% water vapor and the stack temperature is 700 °F (371 °C).

Solution:

(a) Interpolating the values at 500 and 1,100 °F, the specific heat capacities of dry air and water vapor at 700 °F can be estimated as 0.253 and 0.489 Btu/lb-°F, respectively.

(b) Heat capacity of the gas stream = (0.253)(1 − 15%) + (0.489)(15%)

= 0.289 Btu/lb-°F (= 1.21 kJ/kg-K)

2.3.5 Reynolds number

Reynolds number (Re) is a dimensionless number and it is the ratio of inertial force to viscous force of a flowing fluid, as

$$Re = \frac{vL\rho}{\mu} = \left(\frac{vD\rho}{\mu} \; for \; a \; circular \; pipe \right) \qquad (2.12)$$

where L = characteristic system dimension and v, ρ, μ are velocity, density, and viscosity of the fluid, respectively. L is four times the hydraulic radius (R_H). *Hydraulic radius* is the cross-sectional area of the flow divided by the wetted perimeter. For a circular pipe having a full flow, $R_H = \pi r^2 / 2\pi r = r/2 = D/4$ (r = radius and D = diameter). Consequently, its characteristic system dimension is equal to its diameter (i.e., L = D).

When Re < ~2,000, the flow is laminar (i.e., the layers of fluid move in parallel and no mixing of the molecules occurs between the layers). When Re >~4,000, the fluid moves in a turbulent manner. The region lies between Re values of 2,000 and 4,000 is the transition region.

The formula for Reynolds number of particles travelling in air is different from that described in Eq. 2.12. It will be presented in Chapter 6.

2.4 Properties of Compounds of Concern in Air

2.4.1 Vapor pressure

Vapor pressure (P^{vap}) of a compound is the equilibrium pressure of a vapor with its liquid or solid in a closed container. As the temperature increases, the vapor pressure increases. Inter-molecular forces among molecules in a compound determine its vapor pressure. Therefore, some compounds are more volatile than the others. When P = 1 atm, the P^{vap} of water is equal to 1 atm at T = 373K (the boiling point), while it is only 18 mm-Hg @T = 273K. The Antoine equation is commonly used to estimate P^{vap} at a specific temperature as:

$$\ln(P^{vap}) = A - \frac{B}{T+C} \qquad (2.13)$$

where A, B, and C are the Antoine constants, and their values can be readily found from Internet or chemistry handbooks. It should be noted these Antoine constants are empirical constants; and, for a specific compound, their values may be different for different temperature ranges. In addition, the reported values of these constants can differ based on results of different research. One should also pay attention to the units of T and P^{vap} that go with these values (This suggestion applies to all empirical equations).

The Clausius-Clapeyron equation correlates the P^{vap} and the absolute temperature. It assumes that enthalpy of vaporization (ΔH^{vap}) is independent of temperature:

$$\ln \frac{P_1^{vap}}{P_2^{vap}} = - \frac{\Delta H^{vap}}{R} \left[\frac{1}{T_1} - \frac{1}{T_2} \right]. \qquad (2.14)$$

where R = the universal gas constant and T = absolute temperature. Values of ΔH^{vap} can be readily found from Internet or chemistry handbooks. The unit of R should match those of T and ΔH^{vap}.

For an ideal solution, the vapor-liquid equilibrium follows Raoult's law as:

$$P_A = P^{vap} \times x_A \qquad (2.15)$$

where P_A = partial pressure of compound A in the vapor phase and x_A = mole fraction of compound A in the solution. Raoult's law holds only for ideal solutions. In dilute aqueous solutions that are commonly encountered in environmental engineering applications, Henry's law is more suitable (described later).

The *partial pressure* is the pressure that a compound could exert, if all the other gases were not present. This is equivalent to the mole fraction of the compound in the gas phase (y) multiplied by the total pressure (P_{total}):

$$P_A = P_{total} \times y_A \qquad (2.16)$$

The total pressure is the sum of all the partial pressures. For example, the ambient air contains 21% by mole (or by volume) of oxygen. Thus, the mole fraction of oxygen in the air is 0.21; and its partial pressure is 0.21 atm ($= 0.21 \times 1$ atm). It should be noted here that 1% is $10,000 \times 10^{-6}$; that is 10,000 parts per million by volume (ppmV). [Note: The ppmV, or ppm (v/v), is determined by comparing the volume of one constituent with the total volume of the substance. Gas concentrations are always expressed in ppm (v/v) as opposed to the ppm (w/w) format that is often used for liquid or solid. In this book, ppm will be used as an abbreviation of ppmV, or ppm (v/v)]. Thus, the oxygen concentration in our ambient air is about 210,000 ppm. If CO_2 concentration at a location is 400 ppm, then its mole fraction is 0.0004. The corresponding CO_2 concentration in percentage is 0.04% by mole (or by volume). Its partial pressure would be 4×10^{-4} atm or 0.304 mm-Hg ($= 760$ mm-Hg $\times (4 \times 10^{-4})$).

Example 2.5   Estimate partial pressures of a mixed solvent

An above-ground storage tank contains 50% (by wt.) of toluene (C_7H_8, MW $= 92$) and 50% of ethyl benzene (C_8H_{10}; MW $= 102$). Estimate the maximum toluene and ethyl benzene concentrations (in ppm) in the void space of the tank. The ambient temperature is $20\,^\circ C$.

Solution:

(a) At $20\,^\circ$ C, the vapor pressure of toluene $= 22$ mm-Hg and that of ethyl benzene $=$ mm-Hg.

(b) Using a total mass of 100 g as the basis for calculations; for 50% by weight of toluene, its mole fraction

= (moles of toluene) ÷ (moles of toluene) + (moles of ethyl benzene)]

$= (50/92) \div [(50/92) + (50/106)] = 0.535 = 53.5\%$

(c) Using Eq. 2.16, the partial pressure of toluene in the void

$= (22)(0.535) = 11.78$ mm-Hg $= 0.0155$ atm $= 15,500$ ppm

(d) The partial pressure of ethyl benzene in the void

$= (7)(1 - 0.535) = 3.25$ mm-Hg $= 4,300$ ppm

Discussion: The vapor concentrations are those in equilibrium with the solvent. An equilibrium between the liquid and vapor phases can occur in a container or in a stagnant space.

Example 2.6   Using the Antoine Equation to estimate vapor pressure

The empirical constants of the Antoine equation for benzene are: $A = 15.9008$, $B = 2788.51$; $C = -52.36$ for P^{vap} in mm-Hg and T in K (Reid et al., 1987).   Estimate the vapor pressure of benzene at T = 20 and $25\,^{\circ}$C and P = 1 atm using the Antoine equation.

Solution:

(a) Using Eq. 2.13, at $20\,^{\circ}$C (293 K):

$$\ln(P^{vap}) = A - \frac{B}{T+C} = 15.9008 - \frac{2788.51}{(293-52.36)} = 4.322$$

$P^{vap} = \underline{75.3 \text{ mm-Hg}}$

(b) Use Eq. 2.13, at 25 °C (298 K):

$$\ln(P^{vap}) = 15.9008 - \frac{2788.51}{(298-52.36)} = 4.557$$

$P^{vap} = \underline{95.3 \text{ mm-Hg}}$

Discussion: As expected, the vapor pressure of benzene at $25\,^{\circ}$C (95.3 mm-Hg) is larger than that at $20\,^{\circ}$C (75.3 mm-Hg).

Example 2.7   Using the Clausius-Clapeyron equation to estimate vapor pressure

Use the vapor pressure value of benzene at $20\,^{\circ}$C (obtained from Example 2.6) and the Clausius-Clapeyron Equation to estimate its vapor pressure at $25\,^{\circ}$C. Note: enthalpy of vaporization of benzene = 33,830 J/g-mole (Lide, 1992).

Solution:

Using Eq. 2.14,

$$\ln \frac{75.3}{P_2^{vap}} = - \frac{33,840}{8.314} \left[\frac{1}{293} - \frac{1}{298} \right].$$

P^{sat} of benzene at $25\,^{\circ}$C = $\underline{95.1 \text{ mm-Hg}}$.

Discussion: The value (95.1 mm-Hg), derived in this example, by using the Clausius-Clapeyron Equation is essentially the same as the value (95.3 mm-Hg) in Example 2.6 by using the Antoine equation.

2.4.2 Humidity

Humidity is the amount of water vapor present in the air. *Absolute humidity* is the total mass of water vapor in a given volume of air, often in g/m^3; and it is a function of T and P. The amount of water vapor needed to achieve saturation increases with temperature. *Relative humidity* (RH) measures the existing absolute humidity relative to the saturation humidity at that temperature. It can be defined as the ratio of the partial pressure of water vapor (P_{water}) to its vapor pressure in equilibrium with pure water under the same conditions:

$$RH = \left(P_{water} \Big/ P_{water}^{vap} \right) \times 100\% \quad (2.17)$$

Dew point is the temperature when the water vapor starts to condense out of the air. Dry-bulb, wet-bulb, and dew-point temperatures are important parameters to describe the status of air with regards to humidity. The *dry-bulb temperature* is what we measure with a normal thermometer and is basically the temperature of the ambient air. The *wet-bulb temperature* can be measured by wrapping the bulb of the thermometer by a wet fabric. With the cooling of the bulb from adiabatic evaporation of water, the wet-bulb temperature is lower than the corresponding dry-bulb temperature and higher than the dew-point temperature [Note: *"adiabatic"* denotes a process or condition in which heat does not leave or enter the system concerned. In other words, no heat exchange between the system and its surrounding.]. When the *humidity* high, the difference between the wet-bulb and the dry-bulb temperatures would be smaller because less tendency for water to evaporate from the wet fabric. In addition, a smaller difference between the dry-bulb and the dew-point temperatures indicates that the *relative humidity* of the air is high. The relationships among them under standard atmospheric conditions can be retrieved from a *psychrometric chart*. Figure 2.2 shows an example psychrometric chart. Psychrometric charts and tutorials on how to use them are readily available on the Internet.

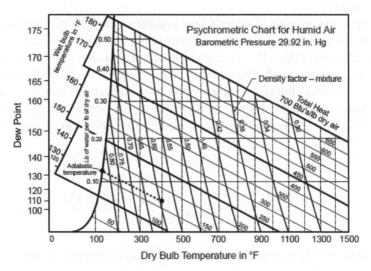

Figure 2.2 - A psychrometric chart (EPA, 2012)

Example 2.8   Calculation of dew-point temperature

The flue gas from a combustion chamber contains 10% of water by volume. Estimate its dew-point temperature (P = 1 atm), using the Antoine equation below (for water vapor between 1 to 100 °C):

$$Log_{10} P \ (in \ mm - Hg) = 8.07131 - \frac{1730.63}{T \ (in \ {}^oC) + 233.426}$$

Solution:

(a) $P_{water} = y_{water} \times P = (10 \ \%)(760) = 76.0$ mm-Hg

(b) $Log_{10}(76.0) = 8.07131 - \frac{1730.63}{T \ (in \ {}^oC) + 233.426}$

   T = 46.1 °C

Discussion:

1. Antonine constants are readily available on the Internet. However, we need to pay attention to the units associated with the parameters since they are empirical constants. For example, "log_{10}" is in the equation above, not the "natural log"; and T is in °C, not in Kelvin.

2. The vapor pressure of water @ 46 °C is 75.7 mm-Hg (from the Internet).

2.4.3 Heating value

In combustion, the combined influent to the combustion chamber should possess sufficient energy to sustain burning. Here are some of the terms commonly used with regards to energy/heat.

Sensible heat (H_s) is the amount of heat needed to cause a change in temperature of a unit mass of a substance without any phase change. It is the multiplication product of specific heat capacity (C_p) and the temperature change (ΔT) as

$$H_s = C_p \times \Delta T \qquad\qquad (2.18)$$

Latent heat, or *heat of vaporization*, (H_v) is the heat/energy required for a unit mass of a substance to transform itself from a liquid state to a vapor state at its boiling point. It involves a change of phases, but the temperature stays the same. For example, water needs to absorb heat to evaporate. The latent heat of water vaporization at its boiling point (i.e., 100 °C at P = 1 atm) is 970.3 Btu/lb, or 2,255 kJ/kg.

Enthalpy (or *heating content*) is the sum of sensible and latent heat present in a substance minus that at reference conditions. The common reference conditions are T = 20 °C and P = 1 atm.

Higher heating value (HHV), also called *gross heating value*, is the total amount of heat released from complete combustion of a fuel at 15.56 °C (60 °F) and its combustion products cooled down to 15.56 °C (60 °F). The *lower heating value* (LHV), or the *net heating value* (NHV), is the HHV minus the latent heat of water formed from combustion. Heating values of gases are often expressed in energy per unit mass or energy per unit volume. Common units for heating value per unit mass are:

- 1 kJ/kg = 0.43 Btu/lb
- 1 Btu/lb = 2.324 kJ/kg

Common units for heating value per unit volume are:

- 1 kJ/m^3 = 0.02685 Btu/ft^3
- 1 Btu/ft^3 = 37.24 kJ/m^3

If the heating value per unit mass of a gas stream is known, its corresponding heating value per unit volume can be readily calculated by multiplying the known value with the density of the gas stream.

27

Organic compounds generally contain higher heating values and they can serve as energy sources for combustion. The higher the organic concentration in a gas stream, the higher the heat content; and, consequently, the less the requirement of auxiliary fuel. Dulong's formula has been commonly used to estimate heat values of fuels or organic substances. If the heating value of a COC is not available, the Dulong's formula can be used:

$$Heating\ Value\ \left(in\ \frac{Btu}{lb}\right) = 145.4\ C + 620\left(H - \frac{O}{8}\right) + 41S \qquad (2.19)$$

$$Heating\ Value\ \left(in\ \frac{kJ}{kg}\right) = 337.9\ C + 1{,}441\left(H - \frac{O}{8}\right) + 95S \qquad (2.20)$$

where C, H, O, and S are the percentages (by weight) of these elements in the compound.

According to U.S. Energy Information Administration (EIA), the average heat content of natural gas in 2015 is 1,037 Btu/ft^3 (38,600 kJ/m^3) and that of the motor gasoline in retail is 120,476 Btu/gallon (33,580 kJ/L). It should be noted that "*therm*" is a common energy unit for natural gas, and 1 therm = 100,000 Btu.

Example 2.9 Estimate the heating value of methane

Use Dulong's formula to estimate the heating value of methane (CH_4) gas.

Solution:

(a) MW of CH_4 = 12 × 1 + 1 × 4 = 16

Weight percentage of C in methane = 12 ÷ 16 = 75%

Weight percentage of H in methane = 4 ÷ 16 = 25%

$$Heating\ Value\ = 145.4\ (75) + 620(25) = 26{,}405\ \frac{Btu}{lb}$$

$$Heating\ Value\ \left(in\ \frac{kJ}{kg}\right) = 337.9(75) + 1{,}441(25) = 61{,}325\ \frac{kJ}{kg}$$

(b) Use Eq. 2.9 to find the density of methane at T = 20 °C (68 °F):

$$\rho = \left(\frac{m}{V}\right) = \frac{P \times MW}{R \times T} = \left(\frac{16}{24.05}\right) = 0.665\ ^{kg}/_{m^3}$$

$$\rho = \left(\frac{m}{V}\right) = \left(\frac{16}{383}\right) = 0.0418\ ^{lb}/_{ft^3}$$

(c) Heating value = (61,325 kJ/kg)(0.665 kg/m^3) = 41,080 kJ/m^3

= (26,405 Btu/lb) × (0.0418 lb/ft^3) = 1,103 lb/ft^3

Discussion:

1. Methane is the principal constituent of natural gas and it represents about 95% of the mixture (Kuo et al., 2015). The heating value of methane calculated from the Dulong's formula, 1,103 Btu/ft^3 or 41,080 kJ/m^3, is very close to the reported average heating value of the natural gas, 1,037 Btu/ft^3 (38,600 kJ/m^3).

2. The weight percentage of C in methane is 75%. A value of 75, not 75%, should be used in the Dulong's formula shown in Eq. 2.19 or 2.20.

3. Methane gas is lighter than air. Its density can also be found by multiplying the air density with the molecular weight ratio of methane and air (i.e., 16/29).

2.5 Interactions among Gas, Liquid, and Solid

2.5.1 Diffusion

Due to molecular diffusion, a COC in air could travel in directions different from that of the air flow. It can also migrate in stagnant air due to diffusion. *Molecular diffusion* is caused by a concentration gradient (i.e., from a high-concentration location to lower concentration locations). Equation below is the 1-D form of the popular Fick's law of diffusion:

$$\frac{\partial G}{\partial t} = D \frac{\partial^2 G}{\partial^2 x} \qquad (2.21)$$

where G = COC concentration, D = diffusion coefficient (length^2/time) and x = distance. It would be harder for a compound of larger MW to diffuse in a fluid and/or when the fluid is more viscous. Generally, diffusion coefficient is inversely proportional to the square root of the MW and inversely proportional to the viscosity of the air as:

$$\frac{D_1}{D_2} = \sqrt{\frac{MW_2}{MW_1}} \qquad (2.22)$$

$$\frac{D_1}{D_2} = \frac{\mu_{air,2}}{\mu_{air,1}} \qquad (2.23)$$

Typical diffusion coefficients of gaseous compounds in air are in the range of 10^{-6} to 10^{-5} m^2/s, while those of the soluble compounds in water are about

10,000 times smaller. For example, the diffusion coefficients of benzene (C_6H_6) @T = 298K & P = 1 atm are 0.92×10^{-5} and 1.02×10^{-9} m^2/s in air and in water, respectively. The diffusivities of particles in air will be discussed in Chapter 6.

2.5.2 Vapor-liquid equilibrium

In addition to gaseous components, air also contains moisture and some liquid droplets. Gaseous components would have a tendency to go across the interfacial boundaries between the air and the liquid droplets to get absorbed or dissolved. On the other hand, dissolved species would also have tendencies to leave the liquid droplets and move into the air. *Absorption* is a common air pollution control process that uses liquid to remove gaseous COCs (more discussion in Chapter 7). Consequently, it is important to understand the interactions between vapor and liquid.

As mentioned earlier, Raoult's law describes the vapor-liquid equilibrium (see Eq. 2.15). In aqueous solutions of lower COC concentrations, which are commonly encountered in environmental engineering applications, Henry's law has been found more applicable. *Henry's law* states that, at equilibrium, the partial pressure of a compound A (P_A) in the gas phase above a liquid is proportional to the concentration of that compound in the liquid (C_A) as

$$P_A = H_A \times C_A \qquad (2.24)$$

where H_A = *Henry's constant* (or the *Henry's law constant*) of compound A. This equation shows a linear relationship between the liquid and gas concentrations. The higher the liquid concentration, the higher the gas concentration will be. A compound with a large Henry's constant means that it prefers to stay in the air. It should be noted that in some air pollution books or literature, Henry's law is written in an opposite way as $C_A = H_A P_A$; the Henry's constant in this format is the inverse of the one shown in Eq. 2.24.

Henry's law can also be expressed in the following form:

$$G = H \times C \qquad (2.25)$$

where C is the COC concentration in the liquid phase and G is the corresponding concentration in the gas phase. Henry's law has been widely used in various disciplines to describe the distribution of a compound between the gas and the liquid phases. The units of the Henry's constant reported in literature vary considerably. The units commonly-encountered include atm/mole fraction, atm/M, M/atm, atm/(mg/L), and may be dimensionless. When inserting a value of Henry's constant into one of the two equations above, it is important to check whether its units match with the two concentrations. Engineers, normally use the units they are familiar with and often have difficulties in performing the necessary unit conversions. Uses of Henry's constant in dimensionless form have been increasing because it is easier (see the design calculations of absorbers in Chapter 7 later). Please be noted that it is not a "(mole fraction)/(mole fraction)" dimensionless unit. The actual meaning of the Henry's constant in this dimensionless format is (concentration in the gas phase)/(concentration in the liquid phase), which can be (mg/L)/(mg/L). To be more precise, it has a unit of (unit volume of liquid)/(unit volume of air). Table 2.3 tabulates conversions among commonly used units for Henry's constant.

Table 2.3 - Unit conversions for Henry's constant
(Kuo & Cordery, 1988)

Desired unit for Henry's constant (H)	Conversion equation
atm/M, or atm·L/mole	$H = H^*RT$
atm·m³/mole	$H = H^*RT/1,000$
M/atm	$H = 1/(H^*RT)$
atm/(mole fraction in liquid), or atm	$H = (H^*RT)[1,000g/W]$
(mole fraction in vapor)/(mole fraction in liquid)	$H = (H^*RT)[1,000g/W]/P$
Note: H^* = Henry's constant in the dimensionless form	
g = specific gravity of the solution (1 for dilute solution)	
W = equivalent MW of solution (18 for dilute aqueous solution)	
R = 0.082 atm/(K)(M)	
T = system temperature in Kelvin	
P = system pressure in atm (usually =1 atm)	
M = solution molarity in (g mol/L)	

The Henry's constant of a compound is the ratio of its vapor pressure and solubility, provided that both are measured at the same temperature, that is

$$H = (Vapor\ Pressure) \div (Solubility) \quad (2.26)$$

This equation implies that the higher the vapor pressure, the larger the Henry's constant is. It also indicates that the lower the solubility, the larger the Henry's constant. For most organic compounds, the vapor pressure increases with temperature and the solubility decreases with temperature. Consequently, Henry's constant, as defined in Eq. 2.26, should increase with temperature.

Example 2.10   Unit Conversions for Henry's Constant

Henry's constant of benzene (C_6H_6) in water at 25 °C is 5.55 atm/M. Convert this value to (a) dimensionless units and (b) units of atm.

Solution:

(a) From Table 2.3,

$$H = H^*RT = 5.55 = H^*(0.082)(273 + 25)$$

$$H^* = \underline{0.227} \text{ (dimensionless)}$$

(b) Also from Table 2.3,

$$H = (H^*RT)[1,000\gamma/W]$$

$$= [(0.227)(0.082)(273+25)][(1,000)(1)/(18)] = \underline{308.3 \text{ atm}}$$

Discussion:

1. Benzene is a VOC of concern and appears in most, if not all, databases of Henry's constant values. It may not be a bad idea to memorize Henry's constant = 0.23 under ambient conditions.

2. To convert the Henry's constant of another COC in the database, just multiply the ratio of the Henry's constants (in any units) of COC and benzene by 0.23. For example, to convert the Henry's constant of methylene chloride (CH_2Cl_2), 2.03 atm/M to dimensionless, multiply the ratio by 0.23 as $[(2.03)/(5.55)] \times (0.23) = 0.084$ (dimensionless).

Example 2.11 Estimate Henry's Constant from Solubility and Vapor Pressure

Vapor pressure of benzene (C_6H_6) and its solubility in water at 298K are 95.2 mm-Hg and 1,780 mg/L, respectively. Estimate its Henry's constant from the given information.

Solution:

(a) $P^{vap} = 95.2/760 = 0.125$ atm

(b) $C = 1,780$ mg/L $= 1.78$ g/L

$= (1.78$ g/L$) \div (78.1$ g/g-mole$) = 0.0228$ mole/L $= 0.0228$M

(c) $H = (0.125$ atm$) \div (0.0228$ M$) = \underline{5.48\ atm/M}$

Discussion: The calculated value, 5.48 atm/M, is essentially the same as the value, 5.55 atm/M, given in Example 2.10.

Example 2.12   Use Henry's Law to estimate equilibrium concentrations

An air sample was taken from the head space of a landfill leachate collection tank. The tetrachloroethylene (PCE, CCl_4) concentration in the air sample was found to be 1,250 ppm. Estimate the PCE concentration in the leachate. Note: T = 298 K, Henry's constant of PCE = 25.9 atm, and MW of PCE = 165.8.

Solution:

(a) $1,250$ ppmV $= 1,250 \times 10^{-6}$ atm $= 1.25 \times 10^{-3}$ atm $= P_A$

(b) From Eq. 2.24,

$P_A = 1.25 \times 10^{-3}$ atm $= H_A \times C_A = (25.9$ atm/M$) \times (C_A)$

(c) $C_A = 4.82 \times 10^{-5}$ M $= (4.82 \times 10^{-5}$ mole/L$)(165.8$ g/mole$)$

$= 8 \times 10^{-3}$ g/L $= \underline{8\ mg/L} = \underline{8\ ppm}$

We can use the dimensionless Henry's constant to solve this problem.

(a) $H = H^* RT = 25.9 = H^*(0.082)(273 + 20)$

$H^* = 1.08$ (dimensionless)

(b) Use Eq. 4.4 to convert ppm to mg/m^3 (see Chapter 4):

$1,250$ ppm $= (1,250)[(165.8/24.05)]$ mg/m^3

$= 8,620$ mg/m$^3 = 8.62$ mg/L

(c) Use Eq. 2.25:

$G = 8.62$ mg/L $= H \times C = 1.08 \times C$

So, $C = \underline{8\ mg/L} = \underline{8\ ppm}$

33

Discussion:

1. These two approaches yield identical results.
2. Henry's constant of PCE is five times larger than that of benzene (1.08 vs. 0.227).
3. A concentration of 8 mg/L of PCE in the leachate is in equilibrium with a gas concentration of 1,250 ppmV.
4. The numeric value of the gas concentration (1,250 ppm) is much higher than that of the corresponding liquid concentration (8 ppm).

2.5.3 Vapor-solid equilibrium

In addition to gaseous components, air also contains particulates. Gaseous components may have tendencies to get attached to the surface of solids. *Adsorption* is the process in which a compound moves across the interfacial boundary from the liquid or gas phase onto the surface of the solid. Here, let us focus on the interaction between the gas and the solid. Adsorption is caused by interactions among three distinct components: (a) *adsorbent*: the solid onto which the target compound is adsorbed/attached onto; (b) *adsorbate*: the target compound that is being adsorbed onto the surface of the solid; and (c) *carrier gas*: the inert part of the gas stream, typically air, from which the target compounds are to be removed. In adsorption, the adsorbate is removed from the carrier gas and taken by the adsorbent. Adsorption is a common air pollution control process which uses solid (e.g., activated carbon) to remove gaseous contaminants. Consequently, it is important to understand the interactions between gas and solid.

For a system where solid and gas phases coexist, an *adsorption isotherm* describes the equilibrium relationship between the gas and solid phases at a constant temperature. The most popular isotherms are the Langmuir isotherm and the Freundlich isotherm; both were derived in early 1900s. The *Langmuir isotherm* has a theoretical basis which assumes a mono-layer coverage of the adsorbent surface by the adsorbate; while the *Freundlich isotherm* is a semi-empirical relationship. For a Langmuir isotherm, the concentration on the solid increases with increasing concentration in the gas phase until a maximum concentration on the solid is reached. The Langmuir isotherm can be expressed as follows:

$$q = q_{max} \frac{KG}{1+KG} \qquad (2.27)$$

where q is the adsorbed concentration on the solid surface, G is the concentration in the gas phase, K is a constant, and q_{max} is the maximum adsorbed concentration.

The Freundlich isotherm is often expressed in the following form:

$$q = KG^{1/n} \qquad (2.28)$$

Both K and $1/n$ are empirical constants. These constants are different for different adsorbates, adsorbents, and carrier gases. For a given compound, the values may also be different for different temperatures and different concentration ranges. When using an isotherm, we should ensure that the units among the parameters and the empirical constants are consistent.

Both isotherms tell the equilibrium relationships between the concentrations of COCs in the gas phase and those adsorbed on the solid. The higher the vapor concentration, the higher the concentration on the solid would be.

Bibliography

Kuo, J. (2014). *Practical Design Calculations for Groundwater and Soil Remediation* (2nd edition), CRC Press, Boca Raton, Florida.

Kuo, J.; Hicks, T.C.; Drake, B.; Chan, T.F. (2015) "Estimation of Methane Emission for California Natural Gas Industry", *J. Air & Waste Management Association* 65(7), 844-55.

Kuo, J.F. and Cordery, S.A. (1988). *Discussion of Monograph for Air Stripping of VOC from Water*, J. Environ. Eng., V. 114, No. 5, p. 1248-50.

Reid, R.C., Prausnitz, J.M. and Poling, B. F (1987). *The Properties of Liquids and Gases*, 4th Edition, McGraw-Hill, Inc., New York, NY.

USEPA (2012). *APTI 427: Combustion Source Evaluation - Student Manual (3rd Edition)*, prepared by ICES Ltd. for Air Pollution Training Institute, United States Environmental Protection Agency, Research Triangle Park, NC 27711.

Exercises

1. At what temperature, is the numerical value on the Celsius scale is the same as that on the Fahrenheit scale?

2. Between 100 pounds and 500 Newtons, which one is a larger force?

3. A diver is 200 feet (61 m) under the sea, what would be the total pressure exerted on the diver (in atm)?

4. Estimate the molecular weight of an air sample containing 2% moisture by volume. Does it have a higher or lower density than dry air?

5. Estimate the specific heat capacity of flue gas from an emission stack. The gas stream contains 15% water vapor and the stack temperature is 300 °C (572 °C).

6. An above-ground storage tank contains 40% (by weight.) of toluene and 60% of ethyl benzene. Estimate the maximum toluene and ethyl benzene concentrations (in ppm) in the void space of the tank. The ambient temperature is 20 °C.

7. The empirical constants of the Antoine equation for benzene are: A = 15.9008, B = 2788.51; C = -52.36 for P_{vap} in mm-Hg and T in K.

    (a) Estimate the vapor pressure of benzene at 303K.

    (b) Use this value and benzene's enthalpy of vaporization to estimate its vapor pressure at 15 °C.

8. The flue gas from a combustion chamber contains 15% of water by volume. Estimate its dew-point temperature (P = 1 atm), using the Antoine equation below (for water vapor between 1 to 100 °C):

$$Log_{10} P \ (in \ mm - Hg) = 8.07131 - \frac{1730.63}{T \ (in \ ^{\circ}C) + 233.426}$$

9. (a) Use Dulong's formula to estimate the heating value of propane (C_3H_8). (b) Compare this heating value with that of methane (from Example 2.9).

10. Henry's constant of toluene in water at 25 °C is 6.7 atm/M. Convert this value to (a) dimensionless units and (b) units of atm.

11. Vapor pressure of toluene and its solubility in water at 298K are 22.0 mm-Hg and 515 mg/L, respectively.

    (a) Estimate its Henry's constant from the given information.

    (b) Compare this result with that in question #10 above.

12. An air sample was taken from the headspace of a landfill leachate collection tank. The methylene chloride concentration in the air sample was found to be 1,000 ppm. Estimate the methylene chloride concentration in the leachate [Note: T = 298 K and Henry's constant of methylene chloride = 2.03 atm/M].

Chapter 3

Regulatory Approaches to Solve Air Pollution Problems

Good air quality cannot be achieved without having good air quality management. Good air quality management starts with establishment of clear and achievable objectives/goal on air quality and has a regulatory framework that can set up implementable regulations and implement them. To evaluate the air quality, air sampling and monitoring are needed. To have acceptable air quality, anthropogenic emissions of air pollutants need to be regulated and reduced or eliminated. This cannot be done without sampling and analysis of pollutant emissions from various sources.

This chapter starts with a brief coverage on the regulatory framework in the U.S. (Section 3.1). Section 3.2 presents main components of the U.S. federal law on air pollution control, the Clean Air Act, including national ambient air quality standards and emission standards for various sources. Section 3.3 describes approaches on ambient air sampling and analysis. Source testing and continuous monitoring on stationary sources are covered in Section 3.4. Section 3.5 presents emission inventories and emission factors. Air pollution control philosophies and strategies are discussed in Section 3.6.

3.1 Regulatory Framework

The Congress passes laws that govern the United States. For the laws related to protection of environment and public health, the Congress has authorized Environmental Protection Agency (EPA), as a regulatory agency, to create and enforce regulations to implement these laws. In addition, a number of Presidential Executive Orders (EOs) play a central role in EPA's activities.

A proposed regulation will first be published for comments in Federal Register (FR), which is the official daily publication for rules, proposed rules, and notices of federal agencies and organizations, as well as EOs and other presidential documents. Once a final decision is issued, the final regulation is then codified and incorporated into U.S. Code of Federal Regulations (CFR). Title 40 (Protection of Environment) of the CFR deals with EPA's mission of protecting human health and the environment. These regulations are mandatory requirements that can apply to individuals, businesses, state or local governments, non-profit institutions, or others. A summary of the U.S. environmental laws and EOs can be found at https://www.epa.gov/laws-regulations/laws-and-executive-orders.

3.2 Clean Air Act

The Clean Air Act (CAA) of 1970 is the federal law that regulates air emissions from stationary and mobile sources. One of its early goals was to set and achieve National Ambient Air Quality Standards (NAAQS) in every state by 1975 to address the public health and welfare risks posed by certain widespread air pollutants and to regulate emissions of hazardous air pollutants (HAPs). The setting of these standards was coupled with directing the states to develop state implementation plans (SIPs), applicable to appropriate industrial sources in the states, so that these standards could be achieved. To set new goals and dates for achieving attainment of NAAQS, the CAA has been amended in 1977 and 1990.

There are eleven titles associated with the Clean Air Act Amendment of 1990 (CAAA of 1990) and they are tabulated in Table 3.1.

Table 3.1 - Titles of CAAA of 1990

I	Nonattainment
II	Mobile Sources/Clean Fuels
III	Hazardous Air Pollutants
IV	Acid Rain
V	Permits
VI	Ozone Deletion and Global Warming
VII - XI	Miscellaneous Research, Enforcement

3.2.1 National ambient air quality standards (NAAQS)
The EPA reviews the NAAQS primary and secondary standards periodically to promulgate revisions, if necessary, to protect public health and welfare. These standards establish the maximum allowable pollutant concentration levels in ambient air. The current primary and secondary standards of the six criteria pollutants (i.e., CO, lead, NO_2, ozone, SO_2, and PM) are tabulated below.

Table 3.2 - NAAQS of CO

Pollutant	Primary/ Secondar	Averaging Time	Level	Form
CO	primary	8 hours	9 ppm	Not to be exceeded more than once per year
		1 hour	35 ppm	

It is interesting to note that CO is the only criteria pollutant that does not have a secondary NAAQS.

Table 3.3 - NAAQS of Pb

Pollutant	Primary/ Secondar	Averaging Time	Level	Form
Lead	primary & secondary	Rolling 3 month	$0.15 \ \mu g/m^3$	Not to be exceeded

Note: In areas designated non-attainment for Pb standard prior to the promulgation of the current (2008) standards, for which implantation to attain or maintain the current standards have not been submitted and approved, the previous standards (1.5 $\mu g/m^3$ as a calendar quarter average) also remain in effect.

Table 3.4 - NAAQS of NO₂

Pollutant	Primary/ Secondar	Averaging Time	Level	Form
NO₂	primary	1 hour	100 ppb	98th percentile of 1-hour daily maximum concentrations, averaged over 3 years
	primary & secondary	1 year	53 ppb	Annual mean

Note: The level of the annual NO_2 standard is 0.053 ppm. It is shown here in terms of ppb for the purposes of clearer comparison to the 1-hr standard level.

Table 3.5 - NAAQS of O₃

Pollutant	Primary/ Secondar	Averaging Time	Level	Form
Ozone	primary & secondary	8 hours	0.070 ppm	Annual fourth-highest daily maximum 8-hour concentration, averaged over 3 years

Note: Final rule signed October 1, 2015, and effective December 28, 2015. The previous (2008) O_3 standards additionally remain in effect in some areas. Revocation of the previous (2008) O_3 standards and transitioning to the current (2015) standards will be addressed in the implementation rule for the current standards.

Table 3.6 - NAAQS of SO$_2$

Pollutant	Primary/ Secondar	Averaging Time	Level	Form
SO$_2$	primary	1 hour	75 ppb	99th percentile of 1-hour daily maximum concentrations, averaged over 3 years
	secondary	3 hours	0.5 ppm	Not to be exceeded more than once per year

Note: The previous SO$_2$ standards (0.14 ppm 24-hour and 0.03 ppm annual) will additionally remain in effect in certain areas: (1) any area for which it is not yet 1 year since the effective date of designation under the current (2010) standards, and (2) any area for which an implementation plan providing for attainment of the current (2010) standard has not been submitted and approved and which is designated nonattainment under the previous SO$_2$ standards or is not meeting the requirements of a SIP call under the previous SO$_2$ standards (40 CFR 50.4(3)). A SIP call is an EPA action requiring a state to resubmit all or part of its State Implementation Plan to demonstrate attainment of the required NAAQS.

The EPA issued standards for PM in 1971; PM$_{10}$ replaced total suspended particulate (TSP) in 1987; and PM$_{2.5}$ standards were added in 1997.

Table 3.7 - NAAQS of PM

Pollutant		Primary/ Secondar	Averaging Time	Level	Form
PM	PM$_{2.5}$	primary	1 year	12.0 µg/m^3	Annual mean, averaged over 3 years
		secondary	1 year	15.0 µg/m^3	Annual mean, averaged over 3 years
		primary & secondary	24 hours	35 µg/m^3	98th percentile, averaged over 3 years
	PM$_{10}$	primary & secondary	24 hours	150 µg/m^3	Not to be exceeded more than once per year on average over 3 years

The NAAQS are enforced by designating areas of attainment and nonattainment with the standards. *Attainment area* is an area that meets the air quality standard for a criteria pollutant. *Nonattainment area* is a geographical area that does not meet one or more of the NAAQS for the criteria pollutants. The states need to revise their SIPs to include measures to bring the nonattainment areas into attainment.

3.2.2 Air quality implementation plans

A SIP is a collection of regulations and documents used by a state, territory, or local air district to reduce air pollution in areas that do not meet NAAQS. Contents of a typical SIP should include (1) state-adopted control measures which consist of either regulations or source-specific requirements, (2) state-submitted "non-regulatory" components, and (3) additional requirements promulgated by EPA. EPA is required to develop a Federal Implementation Plan (FIP) if a state fails to submit a SIP, or if the plan does not fully comply with the NAAQS.

EPA provides a living document, Menu of Control Measures (MCM), that contains existing emission reduction measures as well as relevant information concerning efficiency and cost effectiveness of these measures, https://www.epa.gov/air-quality-implementation-plans/menu-control-measures-naaqs-implementation.

3.2.3 Emission standards for new and modified sources

Emission limitations depend on whether a source is in an attainment or nonattainment area and they can be different for new and existing sources.

Both the EPA and states promulgate emission standards for new and existing stationary sources. *Existing sources* are those constructed before the implementation of the *New Source Performance Standards* (NSPS), which can be found in 40 CFR Part 60. The NSPS is a set of technology-based standards for new and modified stationary sources and it sets emission limitations for industrial source categories. A *modified source* means there is a physical change in operation of the source that will cause an increase in emissions.

New or modified sources located in attainment areas must comply with any applicable NSPS or, in general, apply *Best Available Control Technology* (BACT). BACT is the maximum degree of emission reduction available, taking economic, energy, and environmental factors into consideration. These sources must also comply with Prevention of Significant Deterioration (PSD) provisions.

For new or modified sources located in a nonattainment area must meet the most stringent, technology-based level of control, called *Lowest Achievable Emission Rate* (LAER). Both BACT and LAER are determined on a case-by-case basis, usually by state or local permitting agencies. In determining LAER, the reviewing authority must consider the most stringent emission limitation contained in any SIP and the lowest emission which is achieved by the same or similar type of source. In other words, there is no consideration of economic, energy, and environmental factors in selecting LAER.

3.2.4 Emission standards for existing sources
There are no federal policies for regulating existing stationary sources because they are the primary responsibility of the individual states. For nonattainment areas, *Reasonable Available Control Technologies* (RACTs) are imposed on existing stationary sources. To provide the states guidance on setting RACT emission limits, the EPA has published a series of documents referred to as Control Technique Guidelines (CTGs) and Alternative Control Techniques (ACTs), see https://www3.epa.gov/ttn/atw/ctg_act.html. The EPA also established the RACT/BACT/LAER Clearinghouse, or RBLC, to provide a central data base of air pollution technology information (https://cfpub.epa.gov/RBLC/)

3.2.5 Emission standards for HAPs
Section 112 of the 1990 CAAA addresses emissions of HAPs and requires issuance of technology-based standards for major sources and certain area sources. *Major sources* are defined as a stationary source or group of stationary sources that emit or have the potential of emitting ≥ 10 tons/yr of a HAP or ≥ 25 tons/yr of a combination of HAPs. An "area source" is any stationary source that is not a major source. For major sources, the EPA is required to establish MACT standards. Eight years after the technology-based MACT standards are issued for a source category, the EPA is required to review the standards to determine if any residual risk exists for that source category and, if necessary, revise the standards to address such risk.

3.2.6 Permits
Under EPA's *New Source Review* (NSR) program, a company plans to have a new or modified source must obtain an NSR permit. The NSR permit is a construction permit which requires the company to minimize air pollution emissions by changing the process to prevent air pollution and/or installing air pollution control equipment.

EPA requires facilities that emit pollution into the air to obtain a permit to operate. This permit (known as an "*operating permit*") contains information about how the facility will comply with established emission standards and guidelines. Operating permit is required for any major stationary source. Potential-to-emit limits are determined by the air quality of the geographical region where the facility is located. A permit contains the specific information about how the facility will comply with established emission standards and guidelines set forth by EPA.

42

3.2.7 *Other air pollution control programs*

There are other air pollution control programs established by EPA under the 1990 CAAA, including programs on acid rain, sulfur dioxide emission reduction, nitrogen oxides emission reduction, stratospheric ozone, visibility, and mobile sources. Details of these programs can be found on the EPA websites.

3.3 *Ambient Air Monitoring*

The ultimate goal of air pollution control is to protect humans and the ecosystem from excessive exposures to air pollutants. Implementation of good ambient air monitoring programs, which determine types and concentrations of pollutants in ambient air is needed for development and evaluation of air pollution control strategies.

Air quality at a given location is usually dynamic and complex. It changes spatially and temporally, mainly because of changes in emission rates as well as in meteorology and topography. In addition, physicochemical and photochemical reactions may also occur on these pollutants in the atmosphere. Ambient air monitoring is a systematic, long-term assessment of pollutant levels to determine whether areas are meeting NAAQS. Per requirements of the CAA, the states need to establish and operate ambient monitoring networks and report the collected data to EPA. Since the main purpose is to support human health objectives, monitoring stations are often installed in populated areas, near large point sources, or near sensitive receptors. Technical information on monitoring programs of the U.S., including the networks of monitoring stations, monitoring methods, and QA/QC procedures, can be found at the website of EPA's Ambient Monitoring Technology Information Center (AMTIC) https://www.epa.gov/amtic.https://www.epa.gov/amtic.https://www.epa.gov/amtic.https://www.epa.gov/amtic.

With regards to air monitoring methods, the AMTIC website provides information on air monitoring methods on criteria pollutants, inorganic air toxics, and organic toxics. For criteria pollutants, EPA has established and requires specific monitoring methods to be used for determination of compliance with the NAAQS and they can be found in 40 CFR Part 50 Appendices A through N. The designated *Federal Reference Methods (FRMs)* specify the measurement that must be followed to determine compliance with the primary and secondary NAAQS. A *Federal Equivalent Method* (FEM) may be used if they have been designated to be equivalent to an FRM.

3.3.1 Sample collection for criteria pollutants

The pollutant concentrations in ambient air usually fluctuate. For example, formation of ozone in ambient air is due to photochemical reactions. Its concentrations in the daytime will be higher than those at night. Consequently, averaging times for sampled compounds of concern (COCs) is an important parameter in ambient air sampling. Figure 3.1 illustrates the differences among the real-time, 1-hr, and 4-hr averaging times for gaseous pollutants. A longer averaging time may not truly reflect fluctuations in concentration and dampens peak concentrations. On the other hand, fluctuations in data from continuous air monitoring instruments may make data analysis challenging without comparing time-averaged concentrations. Selection of averaging times depend on the duration needed to collect a meaningful data and the intended use of the collected data. For example, 24-hr averages would be proper to reflect long-term air quality trend; while 1-hr or 8-hr averaging time would be more representative for pollutants that would reach peak levels within short periods of time (EPA, 2003). The averaging times for NAAQS are: 1-hr and 8-hr averages for CO, rolling 3-month average for Pb, 1-hr and 1-yr averages for NO_2, 8-hr for O_3, 24-hr averages for PM_{10} and $PM_{2.5}$ (and 1-yr average for $PM_{2.5}$), and 1-hr and 3-hr averages for SO_2 (see Section 3.2.1 for details).

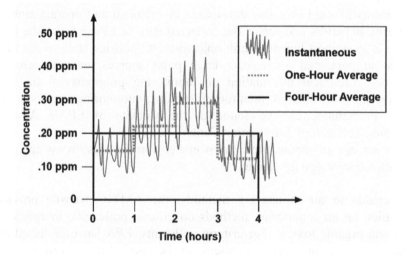

**Figure 3.1 - Averaging times for gaseous pollutants
(Modified from EPA, 2003)**

Filtration and inertial impaction are two of the most important removal mechanisms of particulate control devices (more details in Chapter 6); and they are most commonly used for sampling PM. Inertial impaction is to drive particles to a collection device by deflecting their original paths. Filtration is able to handle large volumetric flow rates. The particulate concentration is readily calculated by using the mass collected and the volume of the air passed through the collection device.

Figure 3.2 illustrates a typical particle size distribution (PSD) and sources of particles in ambient air. The PSD appears to have three peaks at about 0.01, 0.4, and 10 microns (μm, or μ). Most of the larger particles are generated from natural processes as well as mechanical processes. Finer particles are usually anthropogenic (man-made) from combustion sources or are secondary particles. Although large particles (≥15μ) are present in the inhaled air, they would be removed by the cilia found in the nose and throat, and do not present much threat as the finer particles do, because they can enter the lower respiratory system and they are often toxic (e.g., heavy metals and poly-aromatic hydrocarbons).

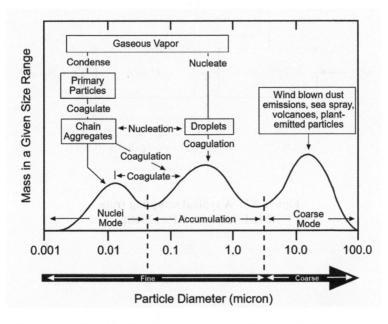

Figure 3.2 - Size distribution and sources of atmospheric particles (EPA, 2003)

Figure 3.3 illustrates a typical sampling train. Air containing pollutants is drawn into the sampling train by an air mover and passes through a sampling collection device. The device removes the pollutants physically and/or chemically from the air stream and holds them for later analyses or analyzes them simultaneously with the collection. A flow meter measures the flow

45

rate so that the total air volume passes through the sampling train can be determined. Due to a relatively small amount of mass, ambient air is drawn to a PM_{10} sampler at a constant flow rate for a relatively long duration. The typical volumetric flow rates are high (e.g., 40 ft^3/min or 1.1 m^3/min); and are often termed "high-volume" samplers. The sampler has an inlet head to remove larger particles through an impaction and gravity-settling chamber before reaching the filters. Figure 3.4 is a schematic of a PM_{10} sampler with an impaction inlet. It should be noted that samplers for $PM_{2.5}$ and PM_{10} are similar, but different. Dichotomous or virtual impact, cascade impactors are often used to quantify $PM_{2.5}$ concentrations in the ambient air (details can be found in EPA (2003)). Typical sampling duration for PM is 24 hours.

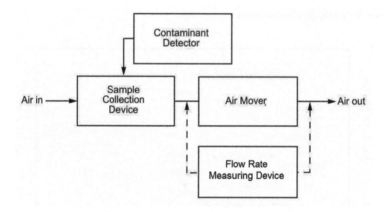

**Figure 3.3 - A typical sampling train
(Modified from EPA, 2008)**

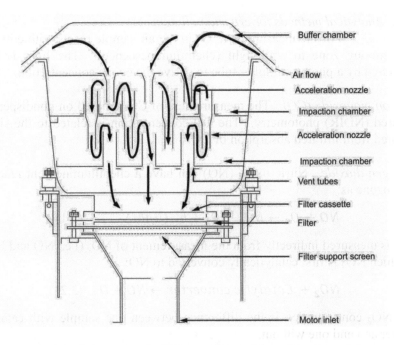

Buffer chamber

Air flow

Acceleration nozzle

Impaction chamber

Acceleration nozzle

Impaction chamber

Vent tubes

Filter cassette

Filter

Filter support screen

Motor inlet

**Figure 3.4 - Schematic of a PM₁₀ sampler with an impaction inlet
(Modified from EPA, 2010)**

Absorption, adsorption, or condensation devices are commonly used to remove gaseous pollutants from gas streams (see Chapter 7 for more details). Adsorption, absorption and condensation are also the operating principles for devices used to collect gaseous samples. The method of choice depends mainly on the physicochemical properties of the pollutants. *Absorption* is transfer of compounds from the bulk of the gas to the air-liquid interface, followed by diffusion of these compounds to the bulk of the liquid. It is essential to get the pollutants dissolved into a solution for subsequent analyses. *Adsorption* is a process whereby gaseous compounds in the air/gas stream pass through a bed of porous material (adsorbent) and these compounds diffuse to the surface of the adsorbent and are retained. Here, the gaseous pollutants will be captured and concentrated on the surface of a solid/adsorbent (e.g., activated carbon). Following the collection, the collected pollutants are purged for subsequent analyses. For condensation or cryogenic trapping, the air stream is drawn through collection chambers with progressively lower temperatures for condensation of the pollutants. The condensate is then subjected to subsequent analyses (EPA, 2008).

3.3.2 Analytical methods for criteria pollutants

Ozone (O_3). For the measurement, O_3 in the air sample reacts with ethylene in a mixing zone to emit light (chemiluminescence). The light is then measured by a photo-multiplier tube to derive the ozone concentration.

Carbon monoxide (CO). The measurement of CO is based on nondispersive infrared (NDIR) photometry. The CO concentration is related to the signal resulted from infrared absorption of CO.

Nitrogen dioxide. Nitric oxide (NO) can have a chemiluminescent reaction with ozone as:

$$NO + O_3 \rightarrow NO_2 + O_2 + hv \, (light) \qquad (3.1)$$

NO_2 is measured indirectly from the measurement of NO_x (i.e., NO and NO_2) in which NO_2 is first catalytically converted to NO:

$$NO_2 + \, Catalytic \, converter \rightarrow NO + O \quad (3.2)$$

The NO_2 concentration is the difference between one sample with catalytic conversion and one without.

Sulfur dioxide (SO_2). The measurement of SO_2 in ambient air is done by the pararosaniline method, a wet chemical method. The air is bubbled through a potassium tetrachloromercurate solution to form a complex (Equation 3.3). Upon addition of pararosaniline, the complex will form a color solution and the SO_2 concentration is then determined in a colorimeter by measuring the light absorbance of the solution.

$$[HgCl_4]^{-2} + 2 \, SO_2 + 2 \, H_2O \rightarrow [Hg(SO_3)]^{-2} + 4 \, Cl^- + 4 \, H^+ \quad (3.3)$$

Particulate matter (PM). Each filter is weighed (after moisture equilibration) before and after sampling to determine the net weight/mass gain due to the captured PM. The total volume of air sampled is determined from the measured air flow rate and the sampling time and corrected to the EPA's reference conditions (25 °C and 1 atm).

Lead. The concentration of lead in the ambient air is actually from measuring the total suspended particulate (TSP) collected by a high-volume sampler. Lead is then acid-extracted from the collected PM and the lead content of the solution is determined by atomic absorption (AA) spectrometry.

Example 3.1   High-volume sampling for PM

Determine the total suspended particulate (TSP) concentration (a) at actual sampling conditions and (b) at standard conditions using the sampling and analysis data below.

- Mass of filter before sampling = 4.4444 g
- Mass of filter after sampling = 4.7777 g
- Ambient temperature = 18 °C
- Ambient pressure = 725 mm-Hg
- Starting flow rate = 1.80 m^3/min
- Ending flow rate = 1.50 m^3/min
- Sampling duration = 24 hours

Solution:

(a) The average actual flow rate (Q_{act}) = (1.80 + 1.50)/2 = 1.65 m^3/min

(b) The actual total sample volume (V_{act}) = (1.65)(1,440) = 2,376 m^3

(c) Mass of TSP collected = 4.7777 – 4.4444 = 0.3333 g

(d) TSP_{actual} = (0.3333)/(2,376) = 1.403 $\times 10^{-4}$ g/m^3 = <u>140 μg/m^3</u>

(e) Using the ideal gas law (Eq. 2.6) to convert the actual volume to volume at standard conditions:

$$V_2 = V_1 \left(\frac{T_2}{T_1}\right)\left(\frac{P_1}{P_2}\right) = (2,376)\left(\frac{298}{291}\right)\left(\frac{725}{760}\right) = 2,321 \ m^3$$

(f) $TSP_{standard\ condition}$ = (0.3333)/(2,321) = 1.436 $\times 10^{-4}$ g/m^3 = <u>144 μg/m^3</u>

Discussion:

1. EPA's standard temperature for ambient air sampling is 298K.

2. The equation used in part (e) is nothing but the gas volume is proportional to the absolute temperature and inversely proportional to the pressure.

3.3.3 Air Quality Index

The air quality index (AQI) is an index for reporting daily ambient air quality. An AQI value is calculated based on ambient air concentrations of O_3, PM, CO, SO_2, and NO_2; five of the six criteria air pollutants, except lead. The range of AQI values is from 0 to 500. The purpose of the AQI is to help the general public to understand the measured ambient air quality to their health. The higher the AQI value, the greater the level of air pollution and the greater health concern. An AQI value of 100 generally corresponds to the NAAQS, which is the level EPA has set to protect public health.

The air quality conditions are considered good, moderate, unhealthy for sensitive groups, unhealthy, very unhealthy, and hazardous when an AQI value is in the range of "0 to 50", "51 to 100", "101 to 150", "151 to 200", "200 to 300", and "301 to 500", respectively. Each category has an assigned color that corresponds to a different level of health concern. Table 3.8 shows the ranges of AQI values and the corresponding levels of health concern as well as the color codes.

Table 3.8 - Air quality index chart

AQI values	Levels of health concern	Color
0 - 50	Good	Green
51 - 100	Moderate	Yellow
101 - 150	Unhealthy for sensitive groups	Orange
151 - 200	Unhealthy	Red
201 to 300	Very unhealthy	Purple
301 - 500	Hazardous	Maroon

3.3.4 Sampling and analysis of ambient HAPs
Ambient air monitoring is to characterize ambient HAPs concentrations and deposition in representative areas. The data can be used to evaluate the dispersion and deposition models and evaluate the effectiveness of HAP reduction strategies (EPA, 2003).

Out of the current 187 listed HAPs, 99 are volatile organic compounds, 49 are semi-volatile organic compounds, 11 are complex organic mixtures, and 28 are non-volatile inorganic compounds. Due to the variety in these HAPs and lack of standard and documented methods, the measurements of ambient HAPs are difficult (EPA, 2003). To address the need to characterize HAP concentrations in ambient air, the EPA has developed and updated compendium of methods. Methods for inorganic toxics are essentially based on "1999 Inorganic (IO) Compendium of Methods" (EPA/625/R-96/01a). For example, Compendium Method IO-5 provides guidance on sampling and analysis for atmospheric mercury. There are 17 compendium methods (Methods TO 1 to 17) for determination of toxic organic (TO) compounds in ambient air (EPA/625/R-96/01b). For example, TO-15 is "Determination of VOCs in Air Collected in Specially-prepared Canisters and Analyzed by GC/MS".

In addition to a sampling chain, grab samples can also be taken by evacuated containers, made of glass, stainless steel or pliable plastics (e.g., Tedlar®, Mylar®, Teflon®, aluminized PVC). A *passivated canister* means that the canister has been passivated to minimize chemical or physical changes to the sample prior to analysis (EPA, 2008). Figure 3.5 shows two examples of containers for "whole air" (grab) sampling.

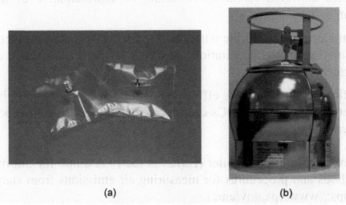

(a)   (b)

Figure 3.5 - Examples of containers for grab sampling:
(a) Tedlar® bags and (b) Summa® canister

3.4 Emission Monitoring and Measurements

To eliminate or reduce emissions of pollutants from human activities into atmosphere is the most effective way to improve the ambient air quality. Measurements of type and quantity of these emissions serve as the basis for implementation of a meaningful air pollution control program. The sampling/monitoring of particulate and gaseous emission from a stationary source is often termed *sources testing* or *source sampling*. Source testing/ sampling is to determine the composition of the effluent and its rate of emission into the atmosphere from a source (e.g., a stack). Typical COCs of source testing include CO, NO_x, SO_2, PM, organics, air toxics, CO_2, and H_2O.

The data/results from a source testing can serve several purposes, such as (i) determining types and emission rates of COCs, (ii) assessing performance of pollution control devices, and (iii) for example, providing feedback to IC engines for tuning to optimize fuel efficiency and to reduce emissions (Kuo & Dow, 2015).

3.4.1 Stationary source emission sampling

To obtain reliable and representative data with regards to the composition of a gas stream and its emission rate, a source test should meet the following requirements:

(a) The gas sampled should be representative of the total or a known portion of stack effluent.

(b) Samples collected for analysis should be representative of the gas stream being sampled.

(c) Volume of the gas sample withdrawn for analysis should be accurately measured so that concentrations of COCs in the stack effluent can be accurately calculated.

(d) The flow rate of the stack effluent should be accurately determined so that emission rates of COCs can be calculated using the corresponding concentrations from (c).

Air Emission Measurement Center (EMC) is the EPA's hub for information on test methods and procedures for measuring air emissions from stationary sources (https://www.epa.gov/emc).

The original eight EPA *Federal Reference Methods* (FRM) were promulgated in 1971 and they were the results of the first New Source Performance Standards (NSPS) authorized in the CAA of 1970 and focused on source categories that emitted large quantities of criteria pollutants. Subsequent FRMs reflect the emission types and characteristics of additional promulgated source categories. The original eight FRMs are:

- FRM 1 - Sample and velocity transverse for stationary source
- FRM 2 - Determination of stack gas velocity and volumetric flow rate
- FRM 3 - Gas analysis of CO_2, oxygen, excess air, and dry MW
- FRM 4 - Determination of moisture content in stack gas
- FRM 5 - Determination of PM emissions
- FRM 6 - Determination of SO_2 emissions
- FRM 7 - Determination of nitrogen oxides emissions
- FRM 8 - Determination of sulfuric acid mist and SO_2 emissions

The first four original FRMs provide the basis for the subsequent methods. Major elements of these methods include sample points and duration, isokinetic sampling, separation of COCs, and calculation of the results.

Since the flow rate is the multiplication product of the velocity and the cross-sectional area, the sampling should start with measurement of velocity. Type "S" Pitot tube is typically used (it comes with large sensing holes to make it less prone to particulate clogging than other types of pitot tubes). The gas is then extracted from sampling points distributed over the cross-section of the stack/duct and the sample duration for each point should be the same (Figure 3.6).

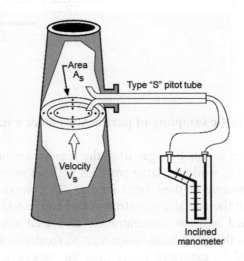

Figure 3.6 - Velocity measurements in stack sampling (EPA, 2003)

For sampling of particles, *isokinetic sampling* [i.e., the sample extraction rate is adjusted so that the velocity of gas into the sample nozzle (v_n) is the same as the velocity of the gas in the stack (v_s)] should be conducted so the extracted sample will have the same concentration as that in the stack (Figure 3.7).

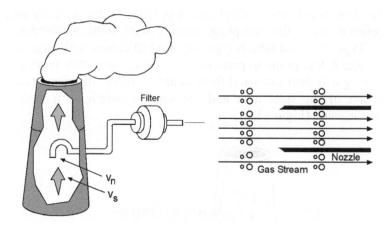

Figure 3.7 - Isokinetic sampling of particulates from a stack (EPA, 2003)

When $v_n > v_s$ (over-isokinetic), gas from the regions around the nozzle will also enter the nozzle, while the large particles will not follow the streamlines into the nozzle because of their large inertia. The measured concentration will be lower than the actual concentration due to the additional volume of stack gas extracted and the concentration is biased with small particles (Figure 3.8a). On the other hand, when $v_n < v_s$ (under-isokinetic), the stack gas volume would be extracted lower than the isokinetic rate. However, larger particles in the streamline immediately outside the nozzle would flow into the nozzle due to their inertia. Consequently, the measured concentration will be higher than the actual concentration due to the smaller volume of stack gas extracted and the concentration is biased with large particles (Figure 3.8b).

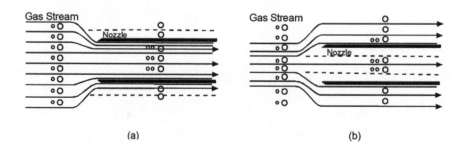

Figure 3.8 - Non-isokinetic conditions: (a) over-isokinetic and (b) under-isokinetic (EPA, 2003)

Figure 3.9 shows a sample train for *Method 5 - Determination of Particulate Emissions from Stationary Sources*. The gas drawn from the stack is passed through a sampling train, a leak-proof series of components to capture the COCs. The first component of the train is a heated probe or a hollow glass; while a pump, the last component, moves the gas through the train. In between, solids are captured on filter paper and water vapor and gaseous pollutants are typically captured by condensation or by bubbling through chilled sealed glass vessels (*impingers*). The volume of the gas sample is measured by a gas meter and then exhausted through an orifice which adjusts the flow rate through the sampling train by a by-pass valve (EPA, 2003).

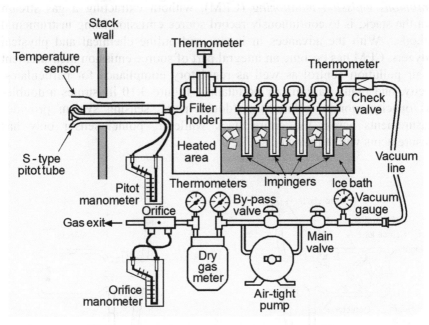

Figure 3.9 - Schematic of a Method 5 sampling train (EPA, 2003)

The mass of particulates collected on the filter paper and the masses of COCs in the solutions of the impingers are then measured and used to determine the concentrations with the measured gas volume. These concentrations are often corrected to the STP and expressed on a dry basis. Typically, the total sampling time for each run lasts about one hour and triplicate runs are conducted (shorter than that of the ambient air sampling, 24 hours (typ.)).

Source testing methods are also promulgated for some HAPs. The EMC developed the numbering system for the source test methods. Method numbers between 1 and 100 are for NSPS; those in the 100 series are for NESHAPs; those in the 200 series are example methods that can be used by the states in their SIPs; and those of the 300 series are for the MACT standards.

3.4.2 Continuous emission monitoring (CEM)

The source emission testing described in the previous section is done on a batch-wise mode, manually, and infrequently. It is time-consuming and labor intensive. In addition, the operators may tend to tune up the process equipment and control device for peak performance before the mandatory/compliance source testing.

The source monitoring systems can be classified into two groups: extractive (e.g., the sampling train that we discussed) and in-situ. In-situ analyzers directly monitor effluent gases at their actual source-level concentrations. *Continuous emission monitoring* (CEM), without extracting a gas stream from the stack, is to continuously record source emission using instrumental methods. With the advances in in-situ and online chemical and physical analyzers, CEM has become an integral part of source emission measurement for air pollution control as well as regulatory compliance for particulates, opacity, and many gaseous air pollutants. Figure 3.10 illustrates a double-pass opacity monitor (Note: the "double-pass" sensing system provides measurements across the entire stack while a "point" sensor only has measurements with a limited distance).

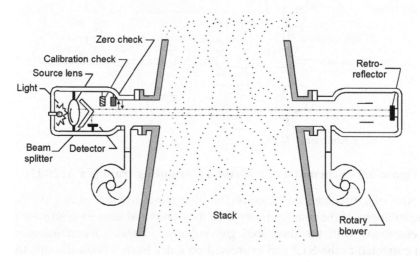

Figure 3.10 - Schematic of a double-pass opacity monitor (EPA, 2003)

3.5 Emission Inventory and Emission Factors

3.5.1 Emission inventory

An emission inventory is a database that contains amounts of air pollutants discharged into the atmosphere from different sources/sectors over a specific time period. National emission inventories can be used to identify significant sources/sectors of air pollutants and use the information as the basis to target necessary regulatory actions. Emission inventories can also be essential input to mathematical models used to estimate air quality. The effect of potential regulatory actions on air quality can be predicted by applying the estimated reduction to the emission inventory data used in the air quality models.

With accumulation of data in emission inventories, emission trends over time can be readily established. Emission inventories include emissions from various sources and sectors and geographical areas. Inventories can also be used to raise public awareness with regards to sources and trends of pollution in the areas that they live in.

The EPA develops and maintains three major emission inventories. The National Emissions Inventory (NEI) is an estimate of air emissions of both criteria pollutants and HAPs from all air emission sources (https://www.epa.gov/air-emissions-inventories). It is updated every three years and is based primary upon information provided by state, local and tribal air agencies for sources under their jurisdictions as well as supplemental data developed by EPA.

The Toxics Release Inventory (TRI) program was created by Section 313 of the Emergency Planning and Community Right-to-Know Act. TRI tracks the management of certain toxic chemicals that may pose a threat to human health and the environment. It is a resource for learning about toxic chemical releases and pollution prevention activities reported by industrial and federal facilities (https://www.epa.gov/toxics-release-inventory-tri-program). EPA also issues an annual report, titled *Inventory of U.S. Greenhouse Gas Emission and Sinks* (https://www.epa.gov/ghgemissions), that tracks U.S. greenhouse gas (GHG) emissions and sinks by source, economic sector, and GHG going back to 1990.

Using the data from NEI, Table 3.9 shows the main sources and emission rates of CO, NO_x, SO_2, PM_{10}, $PM_{2.5}$, and VOCs in the U.S. in 2017. Lots of information can be extracted from this table. For example, CO has the largest emission rate among these pollutants, at ~ 60 million tons. The major emission source for both CO and NO_2 is transportation, while stationary fuel combustion is the main source responsible for SO_2 emissions. Figure 3.11 illustrates the annual emission trends of these air pollutants. As shown, SO_2, NO_x, and CO emissions have been significantly reduced.

Table 3.9 - Source categories and emission trends of several air pollutants (in thousand tons)

Source Category	CO	NO$_x$	SO$_2$	PM$_{10}$	PM$_{2.5}$	VOCs
Stationary fuel combustion	**4,065**	**2,839**	**2,035**	**867**	**749**	**519**
Electric utility	731	1,155	1,385	234	182	38
Industrial	926	1,143	534	286	224	110
Others	2,408	541	116	348	343	372
Industrial and other Processes	**4,001**	**1,282**	**534**	**1,112**	**603**	**7,557**
Chemical & allied product manufacturing	129	47	123	19	14	77
Metals processing	610	70	105	58	44	29
Petroleum & related industries	702	717	104	33	29	3,145
Other industrial processes	584	330	167	677	265	346
Solvent utilization	2	1	0	4	4	3,052
Storage & transport	8	6	3	43	17	675
Waste disposal & recycling	1,967	110	32	277	230	233
Transportation	**32,162**	**6,355**	**96**	**446**	**290**	**3,457**
Highway vehicles	18,893	3,695	27	261	116	1,801
Off-Highway	13,269	2,660	69	185	174	1,656
Miscellaneous	**19,882**	**301**	**150**	**15,727**	**3,701**	**4,699**
Wildfires	10,487	119	71	1,046	886	2,466
Others	9,395	181	79	14,681	2,814	2,232
Total	**60,109**	**10,776**	**2,815**	**18,152**	**5,343**	**16,232**

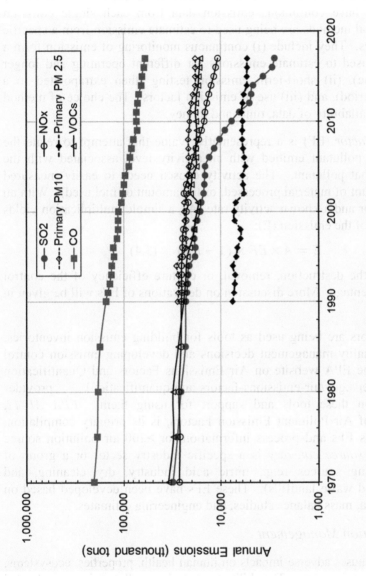

Figure 3.11 - Trends of annual emissions of air pollutants

3.5.2 Emission factors

Emission data are needed to build an emission inventory. It would be impossible to have continuous emission data from each single emission source. Several methods are being used to estimate emission from a specific type of sources. They include (i) continuous monitoring of emission form a source (then used to estimate emissions for different operating and longer averaging time), (ii) short-term emission testing (then extrapolated to a longer time period), and (iii) use of emission factors. The choice of method depends on availability of data, time, and money.

An *emission factor* (EF) is a representative value that attempt to relate the quantity of a pollutant emitted with an activity level associated with the emission of that pollutant. The activity chosen needs to easily measured (e.g., the amount of material processed, or the amount of fuel used). With an emission factor and a known activity rate (A), a simple multiplication yields an estimation of the emission (E):

$$E = A \times EF \times (1 - ER) \quad (3.4)$$

where ER is the destruction, removal, or capture efficiency of the control device in percentage. More discussion on derivations of EFs will be given in Chapter 11.

Emission factors are being used as tools for building emission inventories, guiding air quality management decisions and developing emission control strategies. The EPA website on Air Emissions Factors and Quantification (https://www.epa.gov/air-emissions-factors-and-quantification) provides information on these tools and support for using them. *EPA AP-42*, Compilation of Air Pollutant Emission Factors, is its primary compilation, which contains EFs and process information for >200 air pollution source categories. A *source category* is a specific industry sector or a group of similar emitting sources (e.g., nitric acid industry, dry cleaning, and municipal solid waste landfills). These EFs have been developed based on source test data, mass balance studies, and engineering estimates.

3.6 Air Pollution Management

Air pollution causes adverse impacts on human health, properties, ecosystems, and the global environment. The ability to control air pollution in a coordinated manner is known as *air pollution management*. Good air pollution management typically instills effective air pollution control programs that employ a collection of strategies and tactics to reduce air pollution to achieve the ultimate goal of good air quality. These strategies can differ greatly in their purpose and function. This section first describes the air pollution control philosophies, followed by air pollution control strategies.

3.6.1 Air pollution control philosophies

According to Dr. de Nevers, most air pollution control programs are based on one or a combination of four basic philosophies: emission standards, air quality standards, emission taxes, and cost benefit analysis (de Nevers, 1977).

Emission standards. The basic idea of this philosophy is that some maximum possible or practical degrees of emission control exist. Although this maximum degree of control varies among various classes of emitters, this philosophy assumes that it can be determined individually. If all emitters employ their maximum degrees of control, the cleanest possible air quality would be achieved (de Never, 1977). The NSPS and the NESHAP programs in the U.S., which employ MACT and other standards, are essentially based on this philosophy.

Air quality standards. This philosophy assumes that there are threshold values for all air pollutants. Below the threshold values, these pollutants will not result in damages. This is the basis for EPA to establish the NAAQS from using dose-response and other toxicology data (de Nevers 1977; EPA, 2003).

Emission taxes. This philosophy would impose tax on each emitter based on a published scale related to the emission rates. The tax rate would be established in such a manner so that most major polluters would find it more economical to install pollution control systems to reduce emissions to pay less tax (de Nevers, 1977). Many versions of emission taxes have been proposed, but none have become law in the U.S. (EPA, 2003). Instead, there are some economic incentive programs such as *emission trading*, or *cap and trade*. The cap part is a limit on emissions backed by science, while the trade part is a market for companies to buy or sell emission allowances, which provides an incentive for companies to save money by cutting emissions.

Cost-benefit philosophy. This philosophy assumes that either no threshold values for pollutant exist, or they are low enough (but we cannot afford to have air that clean). It is to select a cost-benefit alternative that best minimize the damages resulting from air pollution in the most cost-effective manner. The cost-benefit analysis is difficult because many uncertain factors and variables need to be analyzed. However, simplified approaches are being used. For example, the EPA must take costs into account when establishing RACT and MACT (de Nevers 1977; EPA 2013).

3.6.2 Air pollution control strategies

Air quality management is a continuous process. Most of the information in this subsection was extracted from EPA website on Air Quality Management Process (https://www.epa.gov/air-quality-management-proces). A control strategy related to air quality is a set of techniques and measures, identified and implemented, to reduce air pollution to attain an air quality goal or standard. Environmental, engineering, and economic factors should be considered in designing an effective air quality control strategy. Examples of environmental factors are ambient air quality conditions, meteorological condition, and locations of the emission sources. Examples of engineering factors are characteristics of pollutants, gas stream and performance of the control system. Examples of economic factors are capital cost, operation and maintenance (O&M) costs, and administrative and enforcement costs.

The old saying, prevention is better than cure, is also applicable to air pollution control. Pollution prevention approaches to prevent, reduce, or eliminate pollution at its source should always be considered. Examples include using less toxic materials/fuels, employing a less-polluting industrial process, and improving the efficiency of the process. To minimize, recycle, reuse the waste generated from air pollution control activities should also be part of the air quality management.

Bibliography

de Nevers (1977). *Air Pollution Control Philosophies*, J. *Air Pollution Control Association*, 27(3), 197-218.

LADCO (2009). *APTI 400: Introduction to Air Toxics - Student Manual*, Lake Michigan Air Directors Consortium (LADCO), Rosemount, Illinois 60018.

USEPA (1986). *Emission Factors for Equipment Leaks of VOC and HAP*, EPA/450/3-86/002, United States Environmental Protection Agency, Research Triangle Park, NC 27711.

USEPA (1992). *APTI 474: Continuous Emissions Monitoring Systems - Student Manual (revised)*, Air Pollution Training Institute, United States Environmental Protection Agency, Research Triangle Park, NC 27711.

USEPA (1999). *APTI 444: Air Pollution Field Enforcement - Student Manual*, Office of Air Quality Planning and Standards, United States Environmental Protection Agency, Research Triangle Park, NC 27711.

USEPA (1999). *APTI 470: Quality Assurance for Air Pollution Measurement Systems - Student Manual*, Air Pollution Training Institute, United States Environmental Protection Agency, Research Triangle Park, NC 27711.

USEPA (2002). *APTI 482: Sources and Control of Volatile Organic Air Pollutants - Student Manual (3rd edition)*, prepared by J.W. Crowder for Air Pollution Training Institute, United States Environmental Protection Agency, Research Triangle Park, NC 27711.

USEPA (2003). *APTI 452: Principles and Practices of Air Pollution Control - Student Manual (3rd edition)*, Air Pollution Training Institute, United States Environmental Protection Agency, Research Triangle Park, NC 27711.

USEPA (2008). *APTI 435: Atmospheric Sampling Course – Student Manual (5th Edition)*, prepared by Tidewater Operations Center of C^2 Technologies for Air Pollution Training Institute, United States Environmental Protection Agency, Research Triangle Park, NC 27711.

USEPA (2009). *APTI 468: Monitoring Compliance Test and Source Test Observation - Course Manual (draft)*, Air Pollution Training Institute, United States Environmental Protection Agency, Research Triangle Park, NC 27711.

USEPA (2010). *APTI 464: Analytical Methods for Air Quality Standards - Student Manual*, Air Pollution Training Institute, United States Environmental Protection Agency, Research Triangle Park, NC 27711.

USEPA (2013). *APTI 450: Source Sampling for Pollutants - Student Guide*, Air Pollution Training Institute, United States Environmental Protection Agency, Research Triangle Park, NC 27711.

Exercise Questions

1. Determine the total suspended particulate (TSP) concentration (a) at actual sampling conditions and (b) at standard conditions using the sampling and analysis data below.
   - Mass of filter before sampling = 3.1234 g
   - Mass of filter after sampling = 5.4321 g
   - Ambient temperature = 22 °C
   - Ambient pressure = 750 mm-Hg
   - Starting flow rate = 2.00 m³/min
   - Ending flow rate = 1.80 m³/min
   - Sampling duration = 24 hours

2. USEPA estimated that CO_2 emission of average passenger cars was 415 g/mile (258 g/km) in 2000. When a car travels at 55 mi/hr (88.5 km/hr), its exhaust flow rate is 250 ft³/min (7.1 m³/min) and the exhaust contains 14% by volume of CO_2 (all based on T = 662 °F (350 °C) and P = 1 atm).

   (a) Compute the CO_2 emission rate, in g/mi or g/km.

   (b) Is this emission rate above or lower than the national average?

Chapter 4

Basic Principles for Engineering Calculations

For a given pollution control device with a known physical configuration, the gas stream flow rate will determine the residence time and flow velocity in this control device. With the mass concentration of an air pollutant in the gas stream, its mass loading rate to the control device can be determined. The residence time, gas flow velocity, and mass loading rate are important design and operating parameters for the control device.

Gas flow rates are often expressed in different ways (e.g., molar, volumetric, and mass) to suit for different applications and operating conditions. Section 4.1 describes how these flow rates are related and illustrates the conversions among them. Similarly, gaseous pollutant concentrations are often expressed in ppm (or ppb), mass concentration, or partial pressure. Section 4.2 illustrates conversions among them. Section 4.3 shows how to determine the residence time and mass loading rate. An emission capture and transport system is an integral part of an air pollution control and removal system. Section 4.4 describes the main components of such a system (e.g., hood, fan, and ductwork) and demonstrates some relevant engineering calculations.

4.1 Flow Rate and Velocity

Process flow rate of a gas stream can be expressed in three different ways: (i) volumetric flow rate (volume/time), (ii) mass flow rate (mass/time), and (iii) molar flow rate (moles/time). They are related; but one may be more suitable than the others for a specific engineering calculation. For example, a volumetric flow rate (Q_v) should be used to size air ductwork or to determine the residence time of gas in a pollution control device. To determine the mass loading rate to a treatment unit, the mass flow rate (Q_{mass}) should be used. For calculations related to chemical reactions, a molar flow rate (Q_{molar}) would typically be more appropriate.

From the ideal gas law, the volume of a gas depends on the temperature (T) and pressure (P) of the system. Consequently, the Q_v of a gas/air stream is a function of T & P. As a gas stream goes through the unit processes/operations having different T's and/or P's in a treatment process train, its Q_v would be changing; while its mass and molar flow rates stay unchanged (assuming no chemical, biological or thermal reactions occurred).

Q_v is typically expressed in cubic meter per minute (m³/min), cubic meter per second (m³/s or cms), or cubic feet per minute (ft³/min or cfm). One common calculation is to convert Q_v between actual and standard conditions.

The ideal gas law can be used to convert Q_v from one set of T and P to another set. Assuming the ideal gas law is valid, conversions between actual Q_v and standard Q_v for a given gas stream can be easily made by using one of the following two formula:

$$\frac{Q_{v,standard}}{Q_{v,actual}} = \left(\frac{T_{standard}}{T_{actual}}\right)\left(\frac{P_{actual}}{P_{standard}}\right) \qquad (4.1)$$

$$\frac{Q_{v,actual}}{Q_{v,standard}} = \left(\frac{T_{actual}}{T_{standard}}\right)\left(\frac{P_{standard}}{P_{actual}}\right) \qquad (4.2)$$

The value of $Q_{v,actual}$ should be larger at a higher T and/or a lower P.

The velocity through a treatment unit or ductwork can be readily determined as the division product of the actual volumetric flow rate ($Q_{v,actual}$) and the cross-sectional area of the flow (A) as:

$$Velocity = \frac{Q_{v,actual}}{A} \qquad (4.3)$$

For example, to calculate the gas flow velocity through an electrostatic precipitator for removal of particulates (Figure 4.1), the influent gas flow rate should be converted to the actual flow rate inside the electrostatic precipitator by taking the system pressure and temperature into consideration, while the cross-sectional area in this case is W×H.

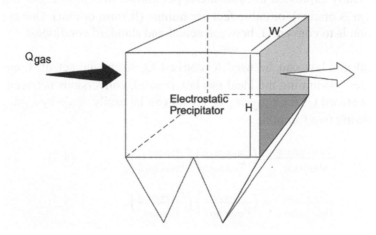

**Figure 4.1 - Example of flow rate and through velocity
(Modified from EPA, 2000)**

The following equations can be used to convert between Q_v, Q_{mass}, and Q_{molar}:

$$Q_{mass} = Q_{molar} \times MW = Q_v \times Gas\ density \qquad (4.4)$$

$$Q_{molar} = Q_v \div Molar\ Volume = Q_{mass} \div MW \qquad (4.5)$$

Example 4.1   Flow rates and velocity

An air stream containing benzene (C_6H_6) is treated by a pollution control system, consisting of a hood, ductwork, activated carbon adsorbers, fan, and exhaust stack. The air flow rate is 200 standard m³/min (7,060 standard ft³/min). The ductwork is operated at T = 50 °C (122 °F) and a static pressure of -8 in-H_2O while the barometric pressure is 29.0 in-Hg. The ductwork dimensions are 0.5 m × 1 m. Find

(a) Q_{mass} and Q_{molar} under the standard conditions,

(b) Q_v, Q_{mass} and Q_{molar} under the actual conditions, and

(c) the air velocity inside the duct.

Solution:

(a) Using Eq. 4.5:

$$Q_{molar} = Q_v \div MV = 200,000 \div 24.05 = 8,316\ \frac{g-mole}{min} = 8.316\ \frac{kg-mole}{min}$$

$$\text{or } Q_{molar} = 7,060 \div 385 = 18.4\ \frac{lb-mole}{min}$$

Using Eq. 4.4:

$$Q_{mass} = Q_{molar} \times MW = \left(8.316 \frac{kg-mole}{min}\right)(29) = 241 \frac{kg}{min} = 530 \frac{lb}{min}$$

(b) $P = P_b + P_g = (29.0 \text{ in-Hg})(407 \text{ in-}H_2O/29.92 \text{ in-Hg}) + (-8 \text{ in-}H_2O)$

$= 394.4 \text{ in-}H_2O$

Using Eq. 4.2:

$$Q_{v,actual} = (200)\left(\frac{273+50}{273+20}\right)\left(\frac{407}{394.4}\right) = 228 \ m^3/min$$

or $\ Q_{v,actual} = (7,060)\left(\frac{460+122}{460+68}\right)\left(\frac{407}{394.4}\right) = 8,031 \ ft^3/min$

Q_{molar} and Q_{mass} will be the same as those in (a).

(c) Using Eq. 4.3:

$$Velocity = Q_{v,actual}/_A = \frac{228 \ m^3/min}{(0.5m \times 0.1m)} = 456 \ m/min = 7.6 \ m/s$$

Discussion:

1. The molar volume of the air is assumed to be 24.05L or 385 ft³ here.

2. 1 atmosphere = 29.92 in-Hg = 407 in-H_2O

3. Typical air velocities in the ducts range from 6 to 33 m/s (or ~ 20 to 100 ft/s), see Section 4.4 for the ranges of transport velocities in ductwork.

4.2 Concentrations

Concentrations of COCs in a gas or air stream are typically expressed in volume concentration (e.g., ppm), mass concentration (e.g., mg/m³), or partial pressures. Conversions between the mass and volume concentrations are often needed. It can be easily done by using one of the two equations below. MW is the molecular weight of the compound and MV is the molar volume of the air at that temperature and pressure:

$$1 \ ppm = \left(\frac{MW}{MV \ (in \ L)}\right) \frac{mg}{m^3} \quad (4.6)$$

$$1 \ ppm = \left(\frac{MW}{MV \ (in \ ft^3)}\right) \times 10^{-6} \frac{lb}{ft^3} \quad (4.7)$$

By the way, 1 lb/ft³ = 1.603×10^7 mg/m³.

Example 4.2 Conversion between volume and gas concentrations

The NAAQS for NO_2 is 0.053 ppm (annual average).

(a) Convert this volume concentration to mass concentration under the standard conditions.

(b) What is the partial pressure of NO_2 in the air (in mm-Hg)?

(c) A dispersion modeling analysis of NO_2 emissions from a source showed a maximum ambient receptor concentration of 220 $\mu g/m^3$. The receptor elevation is 5,000 ft ($P_{barometric}$ = 26.5 in-Hg), and the ambient temperature is 55 °F. Determine the NAAQS value of NO_2 (in $\mu g/m^3$) at this location.

Solution:

(a) Using Eq. 4.6 with MV = 24.05 (or 385 ft^3) and MW = 46,

$$1\ ppm = \left(\frac{46}{24.05}\right)\frac{mg}{m^3} = 1.91\ \frac{mg}{m^3}$$

$$1\ ppm = \left(\frac{46}{385}\right) \times 10^{-6}\frac{lb}{ft^3} = 1.20 \times 10^{-7}\frac{lb}{ft^3}$$

Therefore, 0.053 ppm = (0.053)(1.91) = 0.101 mg/m^3 = 6.33×10^{-9} lb/ft^3.

(b) 0.053 ppm = (0.053 $\times 10^{-6}$) × (760 mm-Hg) = 4.03×10^{-5} mm-Hg.

(c) The molar volume of NO_2 at this location can be determined by using the ideal gas law or

$$\left(\frac{v_2}{v_1}\right) = \left(\frac{T_2}{T_1}\right)\left(\frac{P_1}{P_2}\right)$$

$$MV = (24.05)\left(\frac{460+55}{460+68}\right)\left(\frac{29.92}{23.4}\right) = 30.0L$$

Then, 1 $ppm = \left(\frac{46}{30.0}\right)\frac{mg}{m^3} = 1.53\ \frac{mg}{m^3}$

Thus, 0.053 ppm = (0.053)(1.53) mg/m^3 = 0.081 mg/m^3 = $81\mu g/m^3$.

Discussion

1. Mixed units were used in this example.

2. Similar to what was done in part (b), if the volume at one set of T & P is known, the volume at another temperature and pressure can be calculated by using

$$\left(\frac{v_2}{v_1}\right) = \left(\frac{T_2}{T_1}\right)\left(\frac{P_1}{P_2}\right).$$

4.3 Mass Loading Rate and Residence Time

The mass loading rate (ML) to a treatment unit is the multiplication product of volumetric flow rate (Q_v) and mass concentration (e.g., mg/m^3 or lb/ft^3) as:

$$Mass\ Loading\ (ML) = Q_v \times G \qquad (4.8)$$

It should be noted the values of the volumetric flow rate and the mass concentration used should be under the same conditions (i.e., same T and P).

Similarly, dose would be the multiplication product of the intake volume (V_{intake}) and average mass concentration (G_{avg}) as

$$Dose = V_{intake} \times G_{avg} \qquad (4.9)$$

The residence time (θ) of a treatment unit can be determined by dividing the volume of the unit (V_{unit}) with the actual volumetric flow rate ($Q_{v,actual}$) as

$$\theta = V_{unit} \div Q_{v,actual} \qquad (4.10)$$

Example 4.3: Concentrations, mass Loading, and dose

The primary NAAQS for PM_{10} (24-hr average) is 150 $\mu g/m^3$, and that for SO_2 (1-hr average) is 0.075 ppm. Every time we breathe, we take in about 1 liter of air [Note: for risk assessment, the air intake rate for an adult is typically assumed to be 20 m^3/d]. Assuming the air quality is exactly at these NAAQS values, determine

(a) How many grams of PM_{10} each breath takes in.
(b) The mass concentration of this 0.075 ppm of SO_2 in $\mu g/m^3$ of SO_2.
(c) How many grams of SO_2 each breath takes in.
(d) Assuming particles are spheres with diameter (D_p) = 5 micron and density (ρ) = 2.0 g/cm^3, how many particles each breath takes in.
(e) How many molecules of SO_2 each breath takes in.

Solution:

(a) From Eq. 4.9:

$$Dose = V_{intake} \times G_{avg} = (1L)\left(150\frac{\mu g}{m^3}\right) = 1.5 \times 10^{-7}g$$

(b) From Eq. 4.6:

$$0.075\ ppm = (0.075)\left(\frac{32+16\times 2}{24.05}\right)\frac{mg}{m^3} = 0.2\frac{mg}{m^3} = 200\frac{\mu g}{m^3}$$

(c) $SO_2\ intake = (1L)\left(200\frac{\mu g}{m^3}\right) = 2.0 \times 10^{-7}g$

(d) Mass of one 5μ particle = (volume)(particle density)

$$= \left(\frac{\pi}{6}D_p^3\right)(\rho) = \left[\frac{\pi}{6}(5 \times 10^{-4})^3\right]\left(2\frac{g}{cm^3}\right) = 1.31 \times 10^{-10}g$$

Number of particles = $(1.5 \times 10^{-7}$ g$) \div (1.31 \times 10^{-10}$ g$) = 1,145$

(e) Moles of $SO_2 = (2.0 \times 10^{-7}$ g$) \div (32$ g/mole$) = 6.25 \times 10^{-9}$ moles

Number of SO_2 molecules = $(6.25 \times 10^{-9})(6.022 \times 10^{23}) = 3.76 \times 10^{15}$

Discussion

1. Avogadro number = 6.022×10^{23}.

2. Although each breath does not take in a large mass of PM_{10} or SO_2, great numbers of fine particles can enter the human body.

4.4 Emission Capture and Gas Handling System

Process equipment in a factory may emit pollutants. Many processes are totally enclosed so that all the pollutants are transferred to the air pollution control device. On the other hand, many other processes are open to the surrounding areas. Without being properly captured, the pollutants would disperse directly into the plant air and eventually reach the atmosphere. The main function of hoods is to capture the pollutant air and send it through ductwork, air pollution control (APC) device, and then the stack for discharge. Figure 4.2 illustrates a process flow diagram of an emission capture and gas handling system.

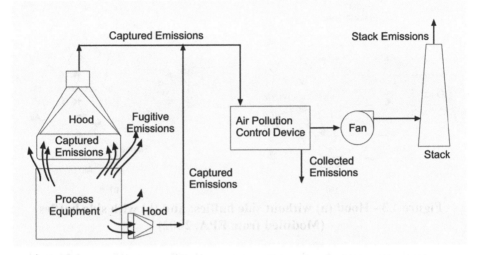

Figure 4.2 - Emission control and gas handling (EPA, 2000)

4.4.1 Hood system

Hoods are used to capture emissions and route them outside or to a control device. Hoods are operated under negative pressure to draw the air in. Once in operation, air from all directions will move toward the hood because it is under negative pressure. Figure 4.3 illustrates how quickly the gas velocity decreases as the distance from the hood increases. The gas velocity at a distance approximately one hood-diameter away from the hood entrance is often <10% of that at the hood entrance. In other words, the influence of hood is insignificant not far away from the entrance. Hoods need to be located as close as possible to emission sources (e.g., process equipment). *Hood capture velocity* (v_{hood}) can be defined as the air velocity at any point in front of the hood or at the hood opening necessary to overcome opposing air currents and to capture the polluted air at that point by pulling it into the hood (EPA, 2012). The required, or the design, hood capture velocity depends on types of pollutants. For example, the design hood capture velocity would be larger for larger and heavier particulates, due to their large inertia, than that for gaseous pollutants.

71

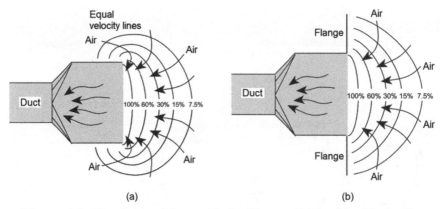

Figure 4.3 - Hood (a) without side baffles; and (b) with side baffles
(Modified from EPA, 2002)

For a freely supported hood without side baffles/flanges, (Figure 4.3(a)), the hood capture velocity at distances within 1.5 times the hood diameter can be estimated by:

$$Q_{hood} = v_{hood}[10x^2 + A] \qquad (4.11)$$

where Q_{hood} = the volumetric air flow rate through the hood, x = distance from the hood entrance to the farthest pollutant release point, v_{hood} = the hood capture velocity at x, and A = the area of the hood opening. As shown in Eq. 4.11, the air flow rate is essentially proportional to x^2; and at $x = 0$, Q_{hood} = $(v_{hood})(A)$. Side baffles/flanges (Figure 4.3(b)) can minimize the flow of clean air into the hood to improve the hood's capture efficiency by either reducing the volumetric hood air flow rate for the same influence distance or increasing the influence distance with the same hood air flow rate as:

$$Q_{hood} = (75\%) \times \{v_{hood}[10x^2 + A]\} \qquad (4.12)$$

Example 4.4   Hood air flow rate

A design hood capture velocity for an air pollutant is 100 m/min for a hood with a diameter of 0.5 m. What would be the hood air flow rate needed to achieve this velocity at 0.5 m and at 1 m away from the hood entrance? If side baffles are added to hood, what will be the required flow rates?

72

Solution:

(a) $A = \pi(0.25)^2 = 0.2 \text{ m}^2$

From Eq. 4.11:

$$Q_{hood} = (100)[10(0.5)^2 + 0.2] = 270 \text{ } m^3/\min \text{ } @ \text{ } x = 0.5 \text{ } m$$

$$Q_{hood} = (100)[10(1.0)^2 + 0.2] = 1,020 \text{ } m^3/\min \text{ } @ \text{ } x = 1.0 \text{ } m$$

(b) With side baffles, the required flow rates will be 75% of the corresponding values in part (a), according to Eq. 4.12.

4.4.2 Ductwork
After the pollutants were captured by the hood and entered the ductwork, a minimum transport velocity needs to be maintained to keep the particles from settling out of the gas stream and depositing onto the bottom of the ductwork. This is practically an important consideration, when handling particle-laden gas streams. Table 4.1 tabulates commonly-recommended transport velocities through ductwork.

Table 4.1 - Transport velocities through ductwork

Type of Pollutant	Transport Velocity	
	(ft/min)	(m/min)
Gases	~1,000 to 2,000	~300 to 600
Light particle loading	~3,000 to 3,500	~900 to 1,070
Normal particle loading	~3,500 to 4,500	~1,070 to 1,370

The proper duct size is a key element in addressing the minimum transport velocity. For a ductwork system with different cross-section sizes, the section with the largest cross-section will be most prone to particle settling due to the smallest transport velocity. However, the section with smaller cross-sections will incur larger pressure drops. The equation below can be used to determine the cross-sectional area of the ductwork (A_{duct}) with a given design transport velocity (v_{duct}) as:

$$A_{duct} = Q_{hood} \div v_{duct} \quad (4.13)$$

4.4.3 Fan system
There are two main types of fans: axial and centrifugal (Figure 4.4). Most fans used in air pollution control systems are centrifugal fans. A centrifugal fan has a fan wheel composed of a number of fan blades mounted around a hub; it can generate high-pressure rises in the gas stream.

73

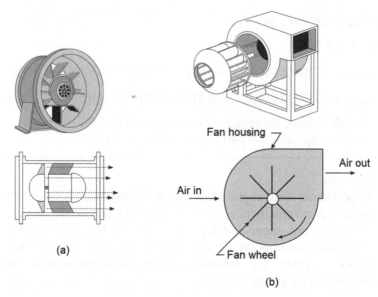

**Figure 4.4 - Types of fans: (a) axial and (b) centrifugal fans
(Modified from EPA, 2002)**

The rotational speed of centrifugal fans is one of the most important operating variables. Most centrifugal fans can operate over a modest range of speeds. There are three fan laws we need to be familiar with. The *first fan law* states that the volumetric gas flow rate is directly proportional to the fan wheel rotational speed as:

$$\frac{Q_2}{Q_1} = \frac{RPM_2}{RPM_1} \qquad (4.14)$$

where RPM is the fan wheel rotational speed in revolutions per minute. Thus, a 20% decrease in fan speed would result in a 20% decrease in the air flow rate through the system.

The *second fan law* states that the fan static pressure rise across the fan (ΔSP) is proportional to the square of the fan speed as:

$$\frac{\Delta SP_2}{\Delta SP_1} = \left(\frac{RPM_2}{RPM_1}\right)^2 \qquad (4.15)$$

The fan's *static pressure rise* must be sufficient to draw the air into the hood to overcome the resistance resulting from flowing though the hood, ductwork, air pollution control systems and stack at the prescribed velocities.

The use of the fan allows the air static pressure to increase from the low level exiting the last APC system to a static pressure slightly above ambient absolute pressure levels (Figure 4.5).

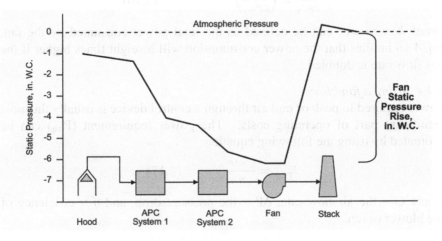

Figure 4.5 - Profile of fan static pressure rise
(Modified from EPA, 2002)

Example 4.5   Fan static pressure rise

The static pressure drop across an emission capture and gas handing system at the fan inlet is -15.0 in-H_2O at a gas flow rate of 200 actual m³/min. Estimate the static pressure rise that the fan needs to provide if the flow rate is increased by 50%.

Solution:

Combining Eqs. 4.14 and 4.15:

$$\frac{\Delta SP_2}{\Delta SP_1} = \left(\frac{RPM_2}{RPM_1}\right)^2 = \left(\frac{Q_2}{Q_1}\right)^2 = \left(\frac{\Delta SP_2}{-15.0}\right) = \left(\frac{300}{200}\right)^2$$

$$\Delta SP_2 = -33.75 \text{ in-}H_2O$$

The fan needs to provide a static pressure rise of at least 33.75 in-H_2O to discharge the air to the atmosphere.

Discussion. It is assumed that the changes in gas density due to the gas flow rate increase is insignificant.

75

The *third fan law* states that the brake horsepower (BHP) is proportional to the cube of the fan rotational speed as:

$$\frac{BHP_2}{BHP_1} = \left(\frac{RPM_2}{RPM_1}\right)^3 \qquad (4.16)$$

Break horsepower can be defined as the total power consumed by the fan. Eq. 4.16 implies that the power consumption will be eight times higher if the gas flow rate is doubled.

4.4.4 Sizing a fan/blower

Power required to push or pull air through a control device is usually the most significant part of operating costs. The power requirement (P_{req}) can be estimated by using the following equation:

$$P_{req} = \frac{Q \times \Delta P}{\eta} \qquad (4.17)$$

where Q = the air flow rate, ΔP = the pressure drop, and η = efficiency of the blower or fan.

Common units for power and their relationships are shown below:

■ 1 horsepower (hp) = 745.7 J/s = 745.7 W

$$= 550 \text{ ft} \cdot lb_f/s = 33,000 \text{ ft} \cdot lb_f/min$$

Example 4.6 Sizing an air blower

A wet Venturi scrubber is to remove fine particles from an air stream of 200 m³/min (7,060 ft³/min) and the pressure drop across the device is 50 cm-H₂O. Assuming the blower efficiency is 85% and the electricity cost is $0.20/kW-hr, estimate the annual electricity cost for a continuous operation of this blower.

Solution:

(a) Convert the unit of the pressure drop:

$$(50 \ cm - H_2O)\left(\frac{1.013 \times 10^5 \frac{N}{m^2}}{1,033 \ cm-H_2O}\right) = 4,903 \ \frac{N}{m^2}$$

(b) Using Eq. 4.17:

$$P_{req} = \frac{\left[(200\frac{m^3}{min})(\frac{min}{60 \ s})\right] \times (4,903\frac{N}{m^2})}{0.85} = 19,230 \frac{N \cdot m}{s} = 19,230 \frac{J}{s} = 19,230 \ W$$

(c) Electricity cost = (19.23 kW)(365 × 24 hrs/yr)($0.20/kW-hr) = $33,700/yr

---

Bibliography

USEPA (1995). *APTI 345: Emission Capture and Gas Handling System Inspection - Student Manual*, Office of Air Quality Planning and Standards, United States Environmental Protection Agency, Research Triangle Park, NC 27711.

USEPA (2002). *APTI 482: Sources and Control of Volatile Organic Air Pollutants - Student Manual (3rd edition)*, prepared by J.W. Crowder for Air Pollution Training Institute, United States Environmental Protection Agency, Research Triangle Park, NC 27711.

USEPA (2002). *EPA Air Pollution Control Cost Manual (6th edition)*, EPA/452/B-02-001, Office of Air Quality Planning and Standards, United States Environmental Protection Agency, Research Triangle Park, NC 27711.

USEPA (2003). *APTI 452: Principles and Practices of Air Pollution Control - Student Manual (3rd edition)*, Air Pollution Training Institute, United States Environmental Protection Agency, Research Triangle Park, NC 27711.

USEPA (2012). *APTI 427: Combustion Source Evaluation - Student Manual*, Office of Air and Radiation, United States Environmental Protection Agency, Research Triangle Park, NC 27711.

Exercise Questions

1. Hazardous waste incinerators are often operated in a negative pressure mode to minimize fugitive emissions. The airflow rate out of a combustion chamber is 10,000 acfm (T = 1600 °F and P = 0.995 atm). What is the mass flow rate (in lb/min)?

2. Hazardous waste incinerators are often operated in a negative pressure mode to minimize fugitive emissions. The airflow rate out of a combustion chamber is 300 actual m^3/min (T = 800 °C and P = 0.995 atm). What is the mass flow rate (in kg/min)?

3. An air stream containing toluene is treated by a pollution control system, consisting of a hood, ductwork, activated carbon adsorbers, fan, and exhaust stack. The air flow rate is 160 standard m^3/min. The ductwork is operated at T = 40 °C and a static pressure of -2.0 kPa while the barometric pressure is 0.98 atm. The ductwork dimensions are 0.4 m × 0.8 m. Find

    (a) Q_{mass} and Q_{molar} under the standard conditions,

    (b) Q_v, Q_{mass} and Q_{molar} under the actual conditions, and

    (c) the air velocity inside the duct.

4. The current 8-hr ozone (O_3) NAAQS is 0.070 ppm. What is the ozone concentration of this standard (in $\mu g/m^3$) on a cold winter day in Denver (barometric pressure = 24.45 in-Hg and temperature = -10 °F)?

5. The current 8-hr ozone (O_3) NAAQS is 0.070 ppm. What is the ozone concentration of this standard (in $\mu g/m^3$) on a cold winter day in Denver (barometric pressure = 0.82 atm and temperature = -20 °C)?

6. Ambient air carbon dioxide (CO_2) concentration has exceeded 400 ppm in many locations. At a location (T = 10 °C, P = 0.95 atm, and [CO_2] = 400 ppm), how many molecules of CO_2 would a local resident inhale in per day (assuming the air intake rate for an adult is 20 m^3/d)?

7. The primary NAAQS for $PM_{2.5}$ (24-hr average) is 35 $\mu g/m^3$, and that for NO_2 (1-hr average) is 100 ppb. Every time we breathe, we take in about 1 liter of air. Assuming the air quality is exactly at these NAAQS values, determine:

    (a) how many grams of $PM_{2.5}$ each are taken in with each breath?

    (b) the mass concentration of NO_2 in $\mu g/m^3$ that is equivalent to 100 ppb of NO_2

    (c) how many grams of NO_2 each breath takes in?

    (d) how many particles each breath takes in (assuming all the particles are spheres with diameter (D_p) = 1 micron and density (ρ) = 1.8 g/cm^3)?

    (e) how many molecules of NO_2 each breath takes in?

8. The primary NAAQS for SO_2 is 75 ppb (1hr-average).

    (a) Convert this volume concentration to mass concentration under the standard conditions.

    (b) What is the partial pressure of SO_2 in the air (in mm-Hg)?

    (c) A dispersion modeling analysis of SO_2 emissions from a source showed a maximum ambient receptor concentration of 60 $\mu g/m^3$. The receptor elevation is 1,000 m ($P_{barometric}$ = 700 mm-Hg), and the ambient temperature is 15 °C. Determine the NAAQS value of SO_2 (in $\mu g/m^3$) at this location.

9. The primary NAQQS standard for SO_2 (1-hr average) is 75 ppb. Industrial representatives claim that the NAAQS for SO_2 is so low that striking a simple wooden match (contains 2.5 mg S) in a modest-sized room (5m × 4m × 3m) at a location $P_{barometric}$ = 0.9 atm and T = 20 °C would cause the SO_2 concentration in the room close to that limit.

Assuming that sulfur is completely burnt into SO_2 and the room is poorly ventilated (for the worst-case scenario),

(a) what would be the molar volume of an ideal gas at $P = 0.9$ atm and $T = 20\,^{\circ}C$?

(b) what would be the maximum concentration of SO_2 in the room in mg/m^3?

(c) what would be the maximum concentration of SO_2 in the room in ppb?

10. The static pressure drop across an emission capture and gas handling system at the fan inlet is -0.2 m-H_2O at a gas flow rate of 200 actual m^3/min. Estimate the static pressure rise that the fan needs to provide if the flow rate is increased by 40%.

11. A wet Venturi scrubber is to remove fine particles from an air stream of 300 m^3/min and the pressure drop across the device is 5.0 kPa. Assuming the blower efficiency is 88% and the electricity cost is $0.18/kW-hr, estimate the annual electricity cost for a continuous operation of this blower.

12. A pollution control device has a pressure drop of 10 inches H_2O at a flow rate of 500 scfm, determine the horsepower required for the blower (efficiency = 85%).

13. The head-loss through a cyclone is 5,000 N/m^2, (16.0 in-H_2O) and the flow rate is 65 ft^3/s.

(a) Estimate the power required if the efficiency of the blower is 85%.

(b) Estimate the annual energy cost if the blower operates continuously at this flow rate and electricity is $0.20/kW-hr.

Chapter 5

Air Pollution Meteorology and Air Pollutant Concentration Models

Air pollution meteorology studies the fate and transport of air pollutants in the atmosphere. Important meteorological parameters include wind speed, wind direction, air turbulence, atmospheric stability, temperature profile, inversion, and mixing height. Topographical factors will also affect the transport and dispersion of the air pollutants. Understanding of air pollution meteorology coupled with air dispersion modeling is essential to air quality management, including determination of air monitoring locations and potential impacts of a new proposed emission source, as well as development of implementation plans.

This chapter starts with a coverage on the temperature and pressure profiles of the troposphere (Section 5.1). Section 5.2 describes horizontal air movement with focuses on horizontal air circulation and wind. Lapse rate, atmospheric stability and inversion layers affect vertical air movement to a greater extent and they are presented in Section 5.3. Section 5.4 explains the relationships between stability class and plume behaviors and the effects of topographical factors. Section 5.5 shows how a plume rise can be estimated. Section 5.6 presents several analytical and numerical air pollution concentration models, including the simplest box model and the commonly-used Gaussian plume dispersion model.

5.1 The troposphere

As mentioned in Chapter 2, the atmosphere is classified into four layers based on its temperature. The troposphere is the part of the atmosphere that we live in. In the troposphere, its temperature and pressure decreases with altitude. According to National Aeronautics and Space Administration (NASA), the temperature (T in °C) and pressure (P in kPa) at an altitude (h in meters) in the troposphere (for h < 11,000 m) can be estimated by the following equations (https://www.grc.nasa.gov/www/K-12/airplane/atmosmet.html):

$$T = 15.04 - (6.49 \times 10^{-3})h \qquad (5.1)$$

$$P = 101.29 \times \left[\frac{T+273.1}{288.08}\right]^{5.256} \qquad (5.2)$$

Clouds and weather generally occur in the troposphere. While large-scale pressure systems control the prevailing meteorology of a region, local meteorological factors include the horizontal air flow (i.e., wind speed and direction) and vertical air movement. These air movements affect the dispersion and transport of compounds of concern (COCs).

Example 5.1   Temperature and velocity profiles in the troposphere

Use the two NASA's equations to answer the questions below:

(a) What is the temperature at the ground level in the equations?

(b) What is the temperature change every kilometer increase in elevation?

(c) What would be the temperature and pressure at the peak of Mountain Everest?

(d) What would be the temperature and pressure at the top of the troposphere?

Solution:

(a) Use $h = 0$ in Eq. 5.1, the temperature of the ground level is 15.04 °C.

(b) As shown in Eq. 5.1, the relationship between T and h is linear and the slope is -0.00649 °C/m (or -6.49 °C/km).

(c) The elevation of Mount Everest is 8,848 m (29,029 ft), so

$$T = 15.04 - (6.49 \times 10^{-3})(8,848) = -42.4 \ °C$$

$$P = 101.29 \times \left[\frac{(-42.4) + 273.1}{288.08} \right]^{5.256} = 31.5 \ kPa = 236 \ mm - Hg$$

(c) The model assumes the top of the troposphere is at 11,000 m:

$$T = 15.04 - (6.49 \times 10^{-3})(11,000) = -56.4 \ °C$$

$$P = 101.29 \times \left[\frac{(-56.4) + 273.1}{288.08} \right]^{5.256} = 22.7 \ kPa = 170.4 \ mm - Hg$$

5.2 Horizontal Movement of Air

5.2.1 Horizontal air circulation

Everything on the earth absorbs, stores, and re-radiates the energy it received from the Sun. Some parts of the Earth heat more readily than the others, and that is known as *differential heating*. Differential heating affects the air above the ground. The air above the warmer spot would be moist, lighter, and buoyant; and, consequently, it forms a low pressure area. One the other hand, the air in a high pressure area is cool and heavy. Wind is created by differences in pressure. Meteorologists use four "scales of motion" to describe the air circulation in the atmosphere. *Microscale* is the smallest of these four and it covers a vertical distance of ≤100 meters and a horizontal distance of ≤ 2 km. Winds in local areas fall into the microscale. The *mesoscale* covers a vertical distance of ≤2 km and a horizontal distance of ≤200 km; and thunderstorms are a mesoscale phenomenon. The *synoptic* scale covers a vertical distance of ≤ 5 km and a horizontal distance of ≤ 1,000 km; and low- and high-pressure systems are synoptic-scale phenomena. The largest scale is the *macroscale* which covers a large portion of the earth.

Since the equator receives more solar irradiation than the poles, the air would be hotter, moist, and buoyant there. It will rise up, while the cold and dense air will move toward the equator. However, the earth rotates and it creates the *Coriolis effect* which makes all moving objects appear to turn to the right. The Coriolis effect makes the air flow situation more complex. As the air flow turns to the right, it divides each hemisphere into three cells. In the Northern Hemisphere, there are polar, mid-latitude, and tropical cells (EPA, 1981).

5.2.2 Wind

Wind is a vector; it has speed and direction. Wind speeds are measured by anemometers. The wind directions are measured by wind vanes. Since the vertical component of wind is small, only horizontal component of wind is considered. Wind direction is the direction where it comes from. Due to the horizontal motion of the wind, a continuous pollutant release is being diluted at its release point. Consequently, pollutant concentration in the plume are inversely proportional to the wind speed. In other words, the stronger the wind, the more dilution would be.

Wind speed is typically recorded at a standard height of 10 m and it is called u_{10}. Wind speed starts from zero and increases to a maximum at some height beyond the influence of ground (e.g., topography and building). To estimate the wind speed at any height of z. a power law relationship is often used:

$$u_z = u_{10} \left(\frac{z}{10}\right)^p \quad (5.3)$$

Where p, the exponent, varies primarily with atmospheric stability and also with terrain. It varies from ~0.07 for unstable conditions to ~0.55 for stable conditions (Turner, 1994).

A *wind rose* is a diagram that depicts the relative frequencies of wind speed and directions at a monitoring location. Wind roses can be used to illustrate seasonal wind patterns as well as local fluctuations by time of day, or others. The length of the spokes of the wind rose in Figure 5.1 indicates wind direction frequency, while the center of the diagram shows the frequency of calm; and each individual segment represents the frequency of the wind speed in the identified range. As shown, the prevailing wind is from southeast. Wind roses can also be used to track or predict dispersion of pollutant from point or area sources. A *pollution rose* can also be constructed to reflect the frequency of measured, or predicted, levels of an air pollutant as a function of wind direction, as illustrated in Figure 5.1 (EPA, 2012).

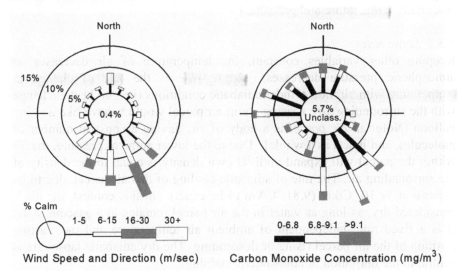

Figure 5.1 - Wind rose and air pollution rose (EPA, 2012)

Example 5.2 Wind velocity profile

The wind velocity recorded at a measurement station, 10 meters above the ground surface, is 3 m/s. What would be the corresponding velocity 40 meters above the ground if the atmosphere was relatively stable (p = 0.5) and if it was relatively unstable (p = 0.1)?

Solution:

(a) For the stable atmosphere (p = 0.5):

$$u_{30} = (3) \left(\frac{40}{10}\right)^{0.5} = 6 \, m/s$$

(b) For the unstable atmosphere (p = 0.1):

$$u_{30} = (3) \left(\frac{40}{10}\right)^{0.1} = 3.45 \, m/s$$

5.3 Vertical Movement of Air

While horizontal movement of air is generally related to wind speed and direction, the forces relate to vertical movement of air is *atmospheric stability*. The mechanisms responsible for the vertical air movement are essentially temperature and pressure.

5.3.1 Lapse rates
Keeping other variables constant, the temperature of air decreases as atmospheric pressure decreases. *Lapse rate* is the rate of change in temperature with altitude. Under adiabatic conditions (i.e., no heat exchange with the surrounding air), a rising warm air parcel would behave like a rising balloon [Note: an *air parcel* is a body of air, having a constant number of molecules, and it acts as a whole]. Due to the lower ambient pressure, the air within the parcel will expand until its own density is equal to the density of the surrounding air. The rate of adiabatic cooling of this air parcel, due to its expansion, is 10 °C/km (9.81 °C/km to be exact). In this context, the air is considered dry, as long as water in the air parcel remains in a gaseous state. It is a fixed rate, independent of ambient air temperature and the starting position of the air parcel rising or descending. The dry adiabatic lapse rate is central to the definition of atmospheric stability.

The actual temperature profile of the ambient air is called the *atmospheric lapse rate*, the *prevailing lapse rate*, or the *environmental lapse rate*. It is the result of complex interactions among meteorological factors. It is usually considered that the temperature decreases with height (similar to the adiabatic lapse rate), but the rate can be larger or smaller than the adiabatic lapse rate. In addition, the ambient temperature sometimes can increase with altitude. In this case, the atmospheric lapse rate will have a sign opposite to that of the adiabatic lapse rate. This layer is termed the *inversion layer*, and it is particularly important to air pollution because it limits vertical air motion. The atmospheric lapse rate is particular important to vertical air movement because the surrounding air temperature determines the extent to which an air parcel rises or falls.

A rising parcel of dry air containing water vapor will continue to cool at the dry adiabatic lapse rate until it reaches the dew point where water starts condensation. Condensation releases latent heat of water within the parcel; consequently, the cooling rate of the air parcel decreases. It is called the *wet adiabatic lapse rate*. The wet adiabatic lapse rate depends on prevailing humidity and varies on locations. However, in the middle latitudes, it is assumed to be about 6 °C/km (EPA, 1982), which is similar to the slope in NASA's equation (Eq. 5.1).

The height that the air parcel's adiabatic lapse rate intersects with the atmospheric lapse rate is the *mixing height* which is the air parcel's maximum level of ascendance. In cases where no intersection occurs (e.g., when the atmospheric lapse rate is constantly greater than or equal to the dry adiabatic rate), the mixing height may extend all the way to the tropopause. The atmosphere below the mixing height is the *mixing layer*. The larger the mixing layer, the greater the volume of air in which air pollutants can be diluted. The height of the mixing layer is not a constant and depends on meteorological conditions.

5.3.2 Atmospheric stability
The degree of atmospheric stability is determined by the temperature difference between an air parcel and its surrounding, which would cause the air parcel to move vertically (i.e., rise or descend). This movement is characterized by four basic conditions: unstable, neutral, stable, and inversion (Figure 5.2).

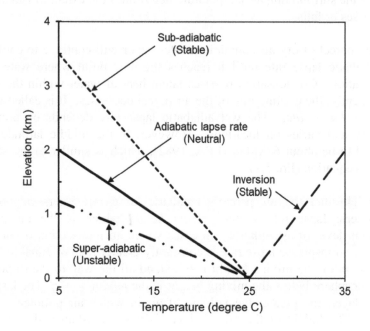

Figure 5.2 - Basic conditions of atmospheric stability

The solid straight line in Figure 5.2 represents the dry adiabatic lapse rate. If the lapse rate of the surrounding atmosphere is greater than the adiabatic rate (i.e., cooling at >10 °C/km), the rising air parcel will continue to be warmer than its surrounding when it moves up. This is the *super-adiabatic rate* (the line below the dry adiabatic rate in Figure 5.2). As shown, the temperature difference between the air parcel and its surrounding increases with height and the buoyancy increases. Consequently, it is an unstable condition. *Unstable conditions* most commonly develop on sunny days with low humidity and wind speed (EPA, 1982).

When the environmental lapse rate is the same as the dry adiabatic lapse rate (in Figure 5.2, these two lapse rates overlap), the atmosphere is in a *neutral condition*. It means that the air movement is neither encouraged nor hindered.

When the surrounding air cools less than 10 °C/km (i.e., the atmospheric lapse rate is less than the adiabatic rate), a rising warm air parcel would cool faster than its surrounding, reaches the temperature of the surrounding at some point, and it will not rise further. This is a *sub-adiabatic lapse rate* (the line above the adiabatic rate in Figure 5.2), in which the air is stable and resists vertical motion of the air parcel. *Stable conditions* most likely occurs on cloudy days with little wind and no strong surface heating, or at night.

As mentioned, the atmospheric temperature decreases as the altitude increases. An *inversion* occurs when the temperature of the atmosphere increases with altitude (see Figure 5.2). The inversion layer acts as a barrier on the vertical movement of air and it would reduce the dispersion of the air pollutants which would result in elevated pollutant concentrations. Inversions are caused by different atmospheric interactions and can persist for different amounts of time.

5.3.3 Types of inversion

By definition, an inversion exists when warmer air overlies cooler air. Inversions can be due to cooling of air at the surface, heating from above, flowing of a layer of warm air over a layer of cold air, or flowing of cooling air under warm air. There are four major types of inversions.

Radiation inversion. It is the most common form of *surface inversion* and occurs when the ground surface cools rapidly. As it cools, so does the layer of air closest to the surface. If the temperature of this air layer becomes cooler than that of the air above, an inversion layer is formed. Radiation inversions usually occur in the late night (*nocturnal radiation inversion*) and early morning under clear skies. During the day, the situation would be reversed. Diurnal cycles of daytime instability and night-time inversions are common. Therefore, the effects of radiation inversions are often short-lived. In locations where radiation inversions are common and relatively close to the ground surface, tall stacks emitting pollutants above the inversion layer can help reduce the ground-level pollutant concentrations (EPA, 1982).

Subsidence inversion. This is almost always associated with high pressure systems in which air descends and flows outward in a clock-wise rotation. As the air descends, the air of higher pressure at lower altitudes compresses and warms at the dry adiabatic rate which is typically faster than the atmospheric lapse rate. The inversion layer will form, and it is elevated at several hundred meters above the surface and during the day. It is called *subsidence inversion,* and it can last a relatively long time (EPA, 1982).

Frontal inversion. The *frontal inversion* is usually associated with cold or hot warm fronts. At the leading edge of either type of the front, the warm air over-rides the cold so that little vertical motion occurs in the cold air closest to the ground. The strength of an inversion depends on the temperature difference between the two air masses. This type of inversion is usually short-lived unless the front becomes stationary (EPA, 1982).

Advective inversion. An *advective inversion* is associated with a horizontal flow of warm air. When warm air flows over a cold surface (e.g., snow cover or extremely cold land), the air closest to the surface will be cooled to form a surface-based inversion. The advective inversion can be terrain-based in which warm air is forced over the top of a cooler air layer. It occurs when warm air from one side of mountain overrides cooler air on the other side of the mountains. Advective inversions are vertically stable, but may have strong winds underneath the inversion layer (EPA, 1982).

5.4 Air Pollution Plume

With regards to air pollution, COCs can enter the atmosphere in different ways (e.g., ground-level, stack, area, or volume source). One method that has received much attention is the release of COCs from stacks (point sources). Although we generally say that dilution is not a solution for pollution control, discharge of COCs into atmosphere is commonly practiced to dilute the strength of a flue gas stream. In this practice, stacks release COCs at elevated locations above the ground surface level so that the COCs can be sufficiently dispersed (diluted) before reaching the ground. The air space that the COCs occupy in the air is often called a *plume*. As the plume travels, it spreads and disperses both horizontally and vertically. For a continuous and constant discharge from a stack, the concentrations at the centers of cross-sections of the plume will decrease as the distance away from the stack increases. In addition, even though the total mass of the COCs of a cross-section stays the same, the size of the plume will generally increase until it hits the ground as the COC concentrations keep on decreasing away from the center due to dispersion. Eventually the plume would hit the ground (to reach the receptors), but at lower COC concentrations than those of the flue gas.

5.4.1 Atmospheric stability and plume behavior
Atmospheric stability and the resulting mixing height have a significant impact on vertical movements of air pollutants in atmosphere. It should be noted that, underneath the inversion layer, surface wind is the dominant factor in dispersing air pollutants. This subsection describes several types of plumes that are characteristic of different stability conditions.

The *looping* plume (Fig. 5.3) results from turbulence caused by overturning of air under highly unstable atmospheric conditions. Although unstable conditions are generally considered to be favorable for pollutant dispersion, high ground-level concentrations (where the receptors are typically located) can occur momentarily if the plume loops downward to the surface.

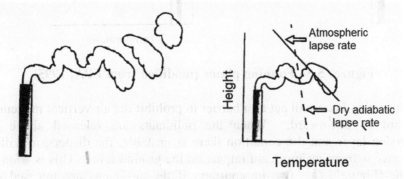

Figurer 5.3 - Looping (modified from EPA, 1995)

The *coning* plume is characteristic of neutral or slightly stable conditions (Figure 5.4) in which vertical air movement is neither encouraged nor hindered. It is likely to occur on cloudy days or mornings of sunny days between the breakup of a radiation inversion and the development of unstable daytime conditions.

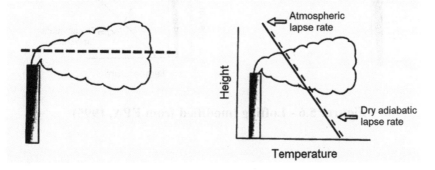

Figurer 5.4 - Coning plume (modified from EPA, 1995)

The *fanning* plume occurs in stable conditions (Figure 5.5). The sub-adiabatic lapse rate discourages the vertical air movement, but not prohibiting horizontal movement. Consequently, the plume may extend a long distance downwind from the release point.

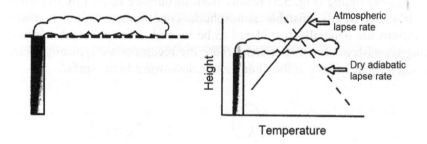

Figurer 5.5 - Fanning plume (modified from EPA, 1995)

An inversion layer will act as a barrier to prohibit the air vertical movement upward or downward. When the pollutants are released above the inversion layer and the condition there is unstable, the dispersion will be effective with an insignificant impact on the ground level. This is a *lofting* plume (Figure 5.6). On the contrary, if the pollutants are released just under the inversion layer, a serious air pollution will develop. The air pollutants cannot travel across the inversion, and they may transport downward to the ground to result in very high ground-level concentrations. This is known as *fumigation* (Figure 5.7).

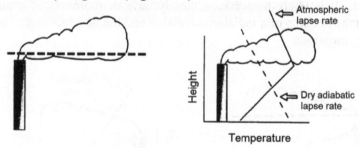

Figurer 5.6 - Lofting (modified from EPA, 1995)

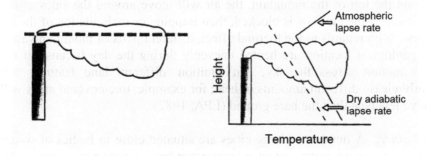

Figurer 5.7 - Fumigation (modified from EPA, 1995)

In addition to the vertical and horizontal movements, turbulence is another mode of air movement. Turbulences consist of circular movements of air and they can be in any directions. The turbulence can be mechanical or buoyant. *Mechanical turbulence* is caused by air movement past an obstruction (e.g., structures and vegetation) or by wind shears (i.e., two adjacent air streams that are moving at different velocities). *Buoyant turbulence* is caused by heating or cooling of air near the ground surface. For example, the heating of the ground surface during a sunny day will create an upward heat flux that heats the air layer adjacent to the ground. It creates an upward-rising thermal stream which creates positive buoyant turbulence (Turner, 1994). Both types of turbulence have impacts on dispersion of pollutants in the air.

5.4.2 Topographical features
The physical characteristics of the earth's surface are often referred to as topographical/terrain features. They can be grouped into four categories: flat, mountain/valley, land/water, and urban (EPA, 1982). They will exert different geometric and thermal effects on air movement.

Flat. Oceans and gently rolling features on land are good examples of this group. The geometric effect of flat terrain is limited to its differences in surface roughness that will result in different wind speed profiles with height. The thermal effect of flat terrain is due to its capacity and rate of absorption and/or release of heat. For example, water does not absorb heat much, while concrete does. The concrete can then release large amounts of heat back to the air (EPA, 1982).

Mountain/valley. Atmospheric dispersion in mountain/valley is obviously more complex than that of flat terrain. It is also called *complex terrain*. Both geometric and thermal effects of mountain/valley terrain are related to the size, shape, and orientations of the features. Air will always find the least resistant path to travel. For example, if there is an temperature inversion layer on the top of the mountain, the air will move around the sides of the mountain. If the air flow is blocked, then trapping or recirculation of the air occurs. With regards to the thermal effect, different parts of mountain/valley in a geological location are heated unevenly during the day because of the Sun's motion across the sky. In addition, different land features will absorb/release different amounts of heat; for example, tree-covered areas will receive less heat than the bare ground (EPA, 1982).

Land/water. A number of large cities are situated close to bodies of water. With regards to geometric effect, a large water body can be considered as flat terrain; while the land has lots of obstructions to air flow and its surface is considered less smooth than that of a large water body. The thermal properties of land and water are drastically different. Water heats and cools relatively slowly, while land and the objects on it will heat and cool at various rates. Water temperatures are relatively stable and follow the seasonal changes, while those of the land vary throughout the day. The warmer daytime temperatures over the land cause the air above it becomes less dense and rise. The cooler air over the water body will be drawn inland, resulting in *sea breeze*. At night, the air over the land cools rapidly, and the air over the water becomes relatively warmer, a land breeze is created. The wind speeds in a land breeze are light, while those in a sea breeze can be quite fast (EPA, 1982). Los Angeles is located in a coastal area, daily sea breeze should help disperse pollutants emitted from automobiles and other emissions. However, the area is also surrounded by high mountains which serve as barriers to horizontal air movement. Instead, the emitted air pollutants can be trapped within the zone. It serves as a good example that topographical groups may have combined effects on air movement and pollutant dispersions.

Urban. Urban areas can exhibit all the characteristics of three topographical features discussed. In addition, urban areas add huge amounts of man-made pollutants to the atmosphere. With buildings and streets, the geometric effect of an urban area is much like complex terrain, or even more complicated. With regards to the thermal effect, buildings absorb and hold lots of heat. The exchange of heat among the buildings are continuous. At night, some of the absorbed heat is transmitted to create a dome over the city, it is called the *heat island effect.*

5.5 Effective Stack Height

For smoke comes out from a stack, we usually notice that the smoke rises above the top of the stack (Figure 5.8). The vertical distance that the plume rises is called the *plume rise* (ΔH), which is calculated as the distance to the imaginary centerline of the plume.

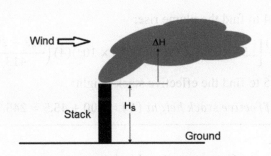

Figurer 5.8 - Plume rise (EPA, 2003)

The extent of the plume rise depends on characteristics of the stack and the effluent gas stream as well as the meteorological conditions around the stack. Holland's formula is often used to estimate the plume rise:

$$\Delta H = \frac{v_s D}{u} \left[1.5 + (2.68 \times 10^{-5})(P)(D) \left(\frac{T_s - T_a}{T_s} \right) \right] \qquad (5.4)$$

where v_s = exit velocity from the stack (m/s), u = wind velocity at the point of release (m/s), P = ambient pressure (Pa), D = diameter of the stack tip, T_s = stack gas temperature (K), and T_a = ambient temperature (K). The calculated value from Eq. 5.4 is often adjusted for atmospheric stability by multiplying it by 1.1 or 1.2 for A and B stability, and 0.8 or 0.9 for D, E, and F stability (Turner, 1994).

The plume centerline distance from the ground is called *effective stack height,* which is the sum of the stack height and the plume rise as:

$$Effective\ stack\ height\ (H) = $$
$$Stack\ height\ (H_s) + Plume\ rise\ (\Delta H) \qquad (5.5)$$

Example 5.3   Effective stack height

Estimate effective stack height of a 500 MW power plant (diameter of the stack = 4 m, stack height = 200 m, ambient temperature = 20 °C, ambient pressure = 1 atm, exit gas velocity = 15 m/s, exit temperature = 140 °C, and stack tip wind speed = 6 m/s).

Solution:

(a) Use Eq. 5.4 to find the plume rise:

$$\Delta H = \frac{(15)(4)}{6}\left[1.5 + (2.68 \times 10^{-5})(1.013 \times 10^{5})(4)\left(\frac{413 - 293}{413}\right)\right] = 46.5\ m$$

(b) Use Eq. 5.5 to find the effective stack height:

$$Effective\ stack\ height\ (H) = 200 + 45.5 = 245.5\ m$$

5.6 Air Pollutant Concentration Models

Air concentration models simulate physical and chemical processes that affect the COCs in the plume. Using characteristics of the emission source (e.g., stack height, exit velocity, concentration, and temperature) and meteorological data (e.g., wind speed, wind direction, and ambient temperature and pressure), the models could estimate the COC concentrations within the plume as well as the maximum concentrations where the plume hit the ground. When the primary air pollutants are not inert, their decay and formation of secondary pollutants through chemical reactions can also be included in the models. These models are important to air quality management because they are widely used to identify potential impacts of a new discharge, evaluate the contribution of an existing discharge to the current situation, and to assist in design effective strategies to improve the ambient air quality, to name a few.

The models can be very simple or very complicated. The simple ones can be solved analytically, while the complicated ones solved numerically.

5.6.1 A simple box model
A box model is a simple model, but it can readily provide an initial estimate of concentration values based on the emission rate and wind speed and direction. The box may represent a city, and the plan area over the city is represented by the length (L) and the width (W), while H is the vertical dimension of the air shed which should be the mixing height.

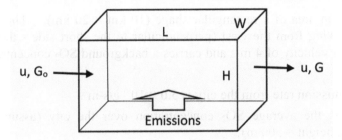

Figure 5.9 - Schematic of a box model in an air shed

It is the simplest model with the following simplified assumptions:

1. Wind is coming in perpendicular to one side of the box at a constant velocity (u), carrying a constant background pollutant concentration (G_o).

2. The city continuously emits the pollutant at a constant flux rate of q.

3. The pollutant will not be destructed or transformed inside the box.

4. The air in the box is completely mixed.

5. No pollutant leaves or enters through the top of the box and the sides that are parallel to the wind direction.

With these assumptions:

Flow rate into and out of the box (Q_{air}) = $[(u)(W \times H)]$ (5.6)

Mass loading rate to the box = $(Q_{air})(G_o)$ = $[(u)(W \times H)](G_o)$ (5.7)

Emission rate = (emission rate/area, q)(area) = (q)(W×L)   (5.8)

Mass outflow rate = $\{[(u)(W \times H)](G_o)\} + [(q)(W \times L)]$ (5.9)

Since the box is well mixed, the effluent pollutant concentration (G) will be the same as the concentration inside the box. It can be found by dividing the mass outflow rate (Eq. 5.9) with the air flow rate (Eq. 5.6) as:

$$G = G_o + \frac{qL}{uH} \qquad (5.10)$$

Example 5.4 The box model

A city has an area of a rectangular shape (10 km × 20 km).   The wind is always blowing from the west (perpendicular to the short side - the 10-km side) with a velocity of 4 m/s and carries a background SO_2 concentration of 10 $\mu g/m^3$.

The SO_2 emission rate from the city is 5.0×10^{-6} g/s-m^2.

(a) What is the average SO_2 concentration over the city (assuming the mixing height = 500 m)?

(b) If the concentration needs to be half of the value in part (a), what does the SO_2 emission rate need to be reduced to?

Solution:

(c) Use Eq. 5.10 to find the effluent concentration:

$$G = G_o + \frac{qL}{uH} = 10 + \frac{(5\frac{\mu g}{s} \cdot m^2)(20{,}000 \ m)}{(4\frac{m}{s})(500 \ m)} = 60 \ \mu g/m^3$$

(d) Use Eq. 5.10 again to find the reduced emission rate:

$$30 = 10 + \frac{(q)(20{,}000 \ m)}{(4\frac{m}{s})(500 \ m)} \rightarrow q = 2 \ \mu g/s \cdot m^2$$

5.6.2 Gaussian plume dispersion model

The Gaussian plume dispersion model is the most commonly-used air concentration model to simulate pollutant dispersion from a point source. Pollutants emitted from the point source travel in the direction of the prevailing wind (the along-the-wind direction).   Small-scale turbulent/eddy motions disperse the pollutants vertically and horizontally in the across-the-plume direction. These dispersion processes determine pollutant concentrations at any location within the plume and at the ground.

Meteorological conditions play an important role in governing the dispersion. If the wind is strong, the pollutants will spread over a large area horizontally, yielding relatively low concentrations. Otherwise, the pollutants will remain in the vicinity of the release point with high concentrations. Atmospheric stability also plays an important role which determines the pollutant dispersion in the vertical direction. The schematic (Figure 5.10) shows the coordinate system used in the model and the shape of the plume.

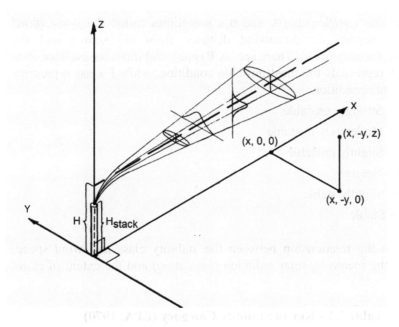

Figure 5.10 - Coordinate system for Gaussian dispersion model (modified from EPA, 1970)

Equation 5.11 describes the three-dimensional concentration generated by a continuous source at a constant emission rate under steady-state meteorological conditions:

$$G(x,y,z) = \frac{Q}{2\pi u \sigma_y \sigma_z} exp\left[-\frac{1}{2}\left(\frac{y^2}{\sigma_y^2}\right)\right]\left\{exp\left[-\frac{1}{2}\left(\frac{(z-H)^2}{\sigma_z^2}\right)\right]\right\} \quad (5.11)$$

Assuming conservation of mass applies and all the pollutants contact with the ground totally reflected, then an imaginary source needs to be added; and the equation becomes:

$$G(x,y,z) =$$
$$\frac{Q}{2\pi u \sigma_y \sigma_z} exp\left[-\frac{1}{2}\left(\frac{y^2}{\sigma_y^2}\right)\right]\left\{exp\left[-\frac{1}{2}\left(\frac{(z-H)^2}{\sigma_z^2}\right)\right] + exp\left[-\frac{1}{2}\left(\frac{(z+H)^2}{\sigma_z^2}\right)\right]\right\}$$
$$(5.12)$$

where G = steady state concentration at a point (x, y, z) within the plume (g/m^3); x = downwind distance along the plume center line (m); y = horizontal distance along the plume center line, Q = emission rate (g/s); u = wind velocity at the release point (m/s), H = effective stack height (in m), and σ_y and σ_z = horizontal and vertical dispersion coefficients (in m).

Both *dispersion coefficients* (σ_y and σ_z), sometimes called *Pasquill-Gifford parameters*, depend on downwind distance from the source and the atmospheric stability class. There are six Pasquill stability classes, from A to F. Class A represents the most unstable condition, while F class represents the most stable condition:

- A = Strongly unstable
- B = Moderately unstable
- C = Slightly unstable
- D = Neutral
- E = Slightly stable
- F = Stable

Table 5.1 is the relationship between the stability class and wind speed, strength of the incoming solar radiation (*insolation*) and the extent of cloud cover.

Table 5.1 - Key to Stability Category (EPA, 1970)

Wind speed (m/s)	Day			Night	
	Incoming solar radiation			Thinly cast (≥ 4/8 cloud)	Clear (≤ 3/8 cloud)
	Strong	Moderate	Slight		
0 - 2	A	A-B	B	-	-
2 - 3	A-B	B	C	E	F
3 - 5	B	B-C	C	D	E
5 - 6	C	C-D	D	D	D
> 6	C	D	D	D	D

Note: 1. Wind speed is measured at 10 m above the ground.

     2. "A-B" means to average the values obtained for each condition.

     3. Neutral class D should be assumed for overcast conditions, day or night.

Figures 5.11 and 5.12 are the Pasquill-Gifford curves for σ_y and σ_z, respectively. For a given downwind distance x and the criteria for stability class (Table 5.1), the two dispersion coefficients can be read from the figures. Two general trends can be readily observed from these two figures: (i) both dispersion coefficients increase with the downwind distance, and (ii) their values increase as the air becomes more unstable.

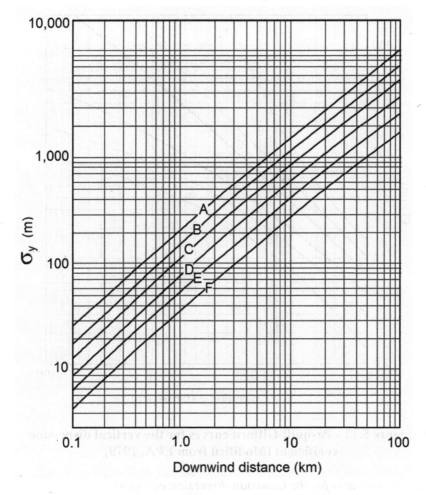

Figure 5.12 - Pasquill-Gifford curves for the horizontal dispersion coefficient (Modified from EPA, 1970)

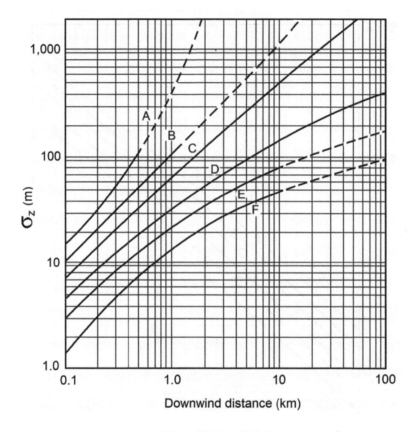

Figure 5.12 - Pasquill-Gifford curves for the vertical dispersion coefficient (Modified from EPA, 1970)

5.6.3 Special cases for the Gaussian dispersion equation
There are several special cases of common interests that can be deduced from the general Gaussian dispersion equation (Eq. 5.12):

(a) Concentrations at the plume centerline (y = 0 and z = H):

$$G(x, 0, H) = \frac{Q}{2\pi u \sigma_y \sigma_z}\left\{1 + exp\left[-\left(\frac{2H^2}{\sigma_z^2}\right)\right]\right\} \qquad (5.13)$$

(b) For receptors at the ground level (z = 0):

$$G(x, y, 0) = \frac{Q}{\pi u \sigma_y \sigma_z} exp\left[-\frac{1}{2}\left(\frac{y^2}{\sigma_y^2}\right)\right]\left\{exp\left[-\frac{1}{2}\left(\frac{H^2}{\sigma_z^2}\right)\right]\right\} \qquad (5.14)$$

(c) For receptors at the ground level and directly beneath the centerline $(z = 0$ and $y = 0)$:

$$G(x, 0, 0) = \frac{Q}{\pi u \sigma_y \sigma_z} \left\{ exp\left[-\frac{1}{2}\left(\frac{H^2}{\sigma_z^2} \right) \right] \right\} \qquad (5.15)$$

(d) For a ground-level release $(H = 0)$, such as fires, explosions, fugitive emissions, or smoldering landfill sites:

$$G(x, y, z) = \frac{Q}{\pi u \sigma_y \sigma_z} exp\left[-\frac{1}{2}\left(\frac{y^2}{\sigma_y^2} \right) \right] \left\{ exp\left[-\frac{1}{2}\left(\frac{z^2}{\sigma_z^2} \right) \right] \right\} \quad (5.16)$$

(e) For a ground-level release and for receptors at ground level $(z = 0,$ and $H = 0)$:

$$G(x, y, 0) = \frac{Q}{\pi u \sigma_y \sigma_z} exp\left[-\frac{1}{2}\left(\frac{y^2}{\sigma_y^2} \right) \right] \qquad (5.17)$$

(f) For a ground-level release and for the maximum ground-level concentration $(y = 0, z = 0,$ and $H = 0)$:

$$G(x, 0, 0) = \frac{Q}{\pi u \sigma_y \sigma_z} \qquad (5.18)$$

Example 5.4   Gaussian plume dispersion model for a stack source

It has been estimated that the SO_2 emission rate from a coal-fired plant is 500 g/s. The physical stack height is 30 m and the plume rise is 6 m. On a bright, sunny day (solar radiation is strong) and the wind speed is 4 m/s. Determine:

(a) the Pasquill stability class
(b) the horizontal dispersion coefficient, at 3 km downwind from the stack
(c) the vertical dispersion coefficient, at 3 km downwind from the stack
(d) the SO_2 concentration at the plume centerline, 3 km from the stack
(e) the maximum SO_2 concentration on the ground, 3 km from the stack
(f) the maximum SO_2 concentration on the ground 3 km from the stack, if the emission is released from the ground-level, not from the tall stack.

Solution:

(a) Stability class: B from Table 5.1 (Sunny day with strong insolation and wind speed of 4 m/s)

(b) Stability class B and $x = 3$ km; from Figure 5.11, $\sigma_y = 420$ m

(c) Stability class B and $x = 3$ km; from Figure 5.12, $\sigma_z = 350$ m

(d) Effective stack height $= 30 + 6 = 36$ m, use Eq. 5.13 to find the centerline SO_2 concentration at $x = 3$ km:

$$G(3,0,36) = \frac{500}{2\pi(4)(420)(350)}\left\{1 + exp\left[-\left(\frac{2(36)^2}{(420)^2}\right)\right]\right\} = 1.07 \times 10^{-3}\frac{g}{m^3}$$

(e) Use Eq. 5.13 to find the maximum. ground SO_2 conc. at $x = 3$ km:

$$G(3,0,0) = \frac{500}{\pi(4)(420)(350)}\left\{exp\left[-\frac{1}{2}\left(\frac{46}{350}\right)^2\right]\right\} = 2.69 \times 10^{-4}\frac{g}{m^3}$$

(f) Use Eq. 5.13 to find the maximum ground SO_2 concentration at $x = 3$ km from the ground-level release:

$$G(3,0,0) = \frac{500}{\pi(4)(420)(350)} = 2.71 \times 10^{-4}\frac{g}{m^3}$$

Discussion: The effective stack height is relatively low, the concentration difference between part (e) and part (f) is insignificant.

5.6.4 Location of the maximum ground-level concentration
Figure 5.13 plots the location of the maximum ground-level concentration downstream from the discharge (x_{max}) versus $(Gu/Q)_{max}$ as a function of the effective stack height (H). $(Gu/Q)_{max}$ can be used to find the maximum ground-level pollutant concentration.

Values of $(Gu/Q)_{max}$ can also be obtained using the equation below (Ranchoux, 1976):

$$\left(\frac{Gu}{Q}\right)_{max} = Exp\{a + b\, lnH + c(\ln H)^2 + d(\ln H)^3\} \qquad (5.19)$$

Table 5.2 - Curve-fitting constants for $(Gu/Q)_{max}$ equation (Ranchoux, 1976)

Stability	Constants			
	a	b	c	d
A	-1.0563	-2.7153	0.1261	0
B	-1.8060	-2.1912	0.0389	0
C	-1.9748	-1.9980	0	0
D	-2.5302	-1.5610	-0.0934	0
E	-1.4496	-2.5910	0.2181	-0.0343
F	-1.0488	-3.2252	0.4977	-0.0765

Example 5.5   Maximum ground concentration from a stack source

For the same stack discharge and the same meteorological conditions, determine the maximum ground-level concentration and the location of this concentration resulted from this discharge.

Solution:

(a) Stability class B and H = 36 m; from Figure 5.13, $(Gu/Q)_{max} = 10^{-4}$ m^{-2}

   $C = (10^{-4}$ m$^{-2})[500$ g/s $\div$ 4 m/s] = 0.0125 g/m^3

(b) Also from Fig. 5.13, x_{max} = 0.26 km

(c) Use Eq. 5.19

$$\left(\frac{Gu}{Q}\right)_{max} = \text{Exp}\{-1.8060 - 2.1912 \ln(36) + 0.0389(\ln(36))^2\}$$

$$= 1.05 \times 10^{-4} m^{-2}$$

Discussion: The $(Gu/Q)_{max}$ values obtained from Figure 5.13 and Eq. 5.19 are essentially the same. However, Ranchoux (1976) did not provide an equation or a figure to find the location of the maximum ground-level concentration.

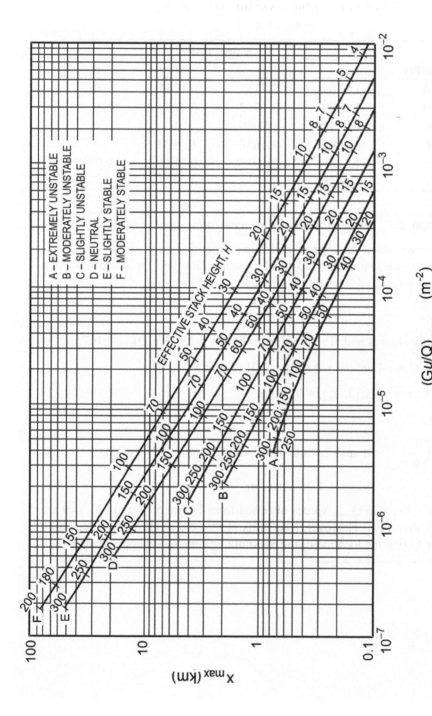

Figure 5.13 - x_{max} versus $(Gu/Q)_{max}$ (NCEES, 2018)

104

5.6.5 EPA models

The EPA has a Support Center for Regulatory Atmospheric Modeling (SCRAM), https://www.epa.gov/scram. Its website provides access to air quality models and other mathematical simulation techniques used in assessing control strategies and source impacts. It covers three types of air quality models: dispersion modeling, photochemical modeling, and receptor modeling.

Dispersion models are typically used in the permitting process to estimate the concentration of pollutants at specified ground-level receptors surrounding an emission source. *Photochemical models* are typically used in regulatory or policy assessments to simulate the impacts from all sources. The impacts are evaluated from the pollutant concentrations and deposition of pollutant, inert or chemically reactive, over large spatial scale. *Receptor models* are observational techniques to identify the presence of and to quantify contribution of emission sources to receptor concentrations. Chemical and physical characteristics of gases and particles measured at the sources and receptor are the input to this type of models.

Wind, atmospheric stability, plume rise and topographical features interact in complex ways to cause the transport and dispersion of air pollutants in the atmosphere. Air quality dispersion modeling uses mathematical formulations to characterize the atmospheric processes that affect the dispersion of a pollutant emitted from a source. Using the source emission data (e.g., emission rate and stack height) relevant meteorological inputs (e.g., temperature, wind direction, wind speed, cloud cover, mixing height), a dispersion model can be used to predict the pollutant concentration at selected downwind receptor locations. EPA's *refined/recommended models* are air quality models listed in Appendix W of 40 CFR Part 51 and they are required to be used for State Implementation Plan (SIP) revisions for existing sources and National Source Review (NSR) and Prevention of Significant Deterioration (PSD) programs. *Screening tools* are models that are often applied to determine if refined modeling is needed. *Alternative models* are those not listed in Appendix W; however, they can be used in regulatory applications with case-by-case justification to the reviewing authority.

Assessment Population Exposure Model (ASPEN) and Industrial Source Complex (ISC) Model are frequently used by EPA in permitting processes for evaluating environmental health impacts. ASPEN is basically a screening tool and it calculates ambient air concentration levels based on meteorology, chemistry, and emission rates of air toxics. ISC model is a steady-state Gaussian plume model that can be used to estimate air pollutant concentrations from sources associated with industrial source complex (EPA, 2003).

5.7 Summary

1. Meteorology plays an important role in dispersion and transport of air pollutants.
2. Dispersion of air pollutants depend on emission characteristics (e.g., emission rate, temperature, exit velocity, and effective stack height), meteorological conditions (e.g., temperature, pressure, stability class), and topographical features.
3. Effective stack height is the sum of physical stack height and plume rise.
4. Air quality models use analytical or numerical techniques to simulate physical and chemical processes that affect the dispersion and reaction of air pollutants in the atmosphere.
5. A box model is the simplest air quality model to assess the effect of emissions on ambient air quality.
6. A Gaussian dispersion equation is commonly used to estimate a 3-D concentration profile generated by a continuous source at a constant emission rate under stationary meteorological conditions.

Bibliography

de Nevers, N (2000). *Air Pollution Control Engineering (2nd edition)*, McGraw-Hill Companies, Inc.

LADCO (2009). *APTI 400: Introduction to Air Toxics - Student Manual*, Lake Michigan Air Directors Consortium (LADCO), Rosemount, Illinois 60018.

NCEES (2018). *FE Reference Handbook (9.5 version for computer-based testing)*, National Council of Examiners for Engineering and Surveying (NCEES), Seneca, SC 29678 (downloadable at https://necees.org/)

Ranchoux, R.J.P. (1976). *Determination of Maximum Ground Level Concentration, J. Air Pollution Control Assoc.* 26(11), 1088-9.

Turner, D.B. (1994). *Workbook of Atmospheric Dispersion Estimates – An Introduction to Dispersion Modeling (2nd edition)*, Lewis Publisher, Boca Raton, Fl.

USEPA (1970). *Workbook of Atmospheric Dispersion Estimates*, prepared by Turner, D. B. for Office of Air Programs, United States Environmental Protection Agency, Research Triangle Park, NC 27711.

USEPA (1975). *APTI 411: Air Pollution Meteorology*, Air Pollution Training Institute, United States Environmental Protection Agency, Research Triangle Park, NC 27711.

USEPA (1980). *Air Quality Modeling - How It Is and How It Is Used*, Office of Air Quality, United States Environmental Protection Agency, Research Triangle Park, NC 27711.

USEPA (1981). *APTI 411: Air Pollution Meteorology - Student Workbook*, EPA 450/2-81-014, Air Pollution Training Institute, United States Environmental Protection Agency, Research Triangle Park, NC 27711.

USEPA (1981). *APTI SI 422: Air Pollution Control Orientation Course - Unit 3 Air Pollution Meteorology (3rd edition)*, EPA 450/2-81-017C, prepared by Northrop Services, Inc. for United States Environmental Protection Agency, Research Triangle Park, NC 27711.

USEPA (1982). *APTI SI 409: Basic Air Pollution Meteorology*, EPA 450/2-82-009, prepared by Northrop Services, Inc. for United States Environmental Protection Agency, Research Triangle Park, NC 27711.

USEPA (2003). *APTI 452: Principles and Practices of Air Pollution Control - Student Manual (3rd edition)*, Air Pollution Training Institute, United States Environmental Protection Agency, Research Triangle Park, NC 27711.

USEPA (2014). *APTI 423: Air Pollution Dispersion Models*, Air Pollution Training Institute, United States Environmental Protection Agency, Research Triangle Park, NC 27711.

Exercise Questions

1. A new city is to be built downwind of an existing city. The planner has decided that the future city is composed of two rectangular zones of identical size (5 km $\times$ 2 km, each). One will be residential and the other industrial. The wind is always blowing from the west carrying a background concentration of 2 mg/m^3. There are lots of lands available and the two blocks can be put in any orientation, but they have to have one adjacent side to each other (long to long, short to long, or short-short) and two sides of each block are perpendicular to the wind.

   - CO emission rate from the industrial zone = 400 g/s•km^2
   - CO emission rate from the residential zone = 100 g/s•km^2
   - Wind speed = 2 m/s
   - Mixing height = 250 m

You are hired as the environmental consult to use the Box Model to find the best and the worst orientations of these two zones that would give the lowest CO concentration and the highest CO concentration at the residential area.

(a) Show these two orientations [Note: no calculations needed, just show the orientations of these two blocks with the direction of the wind]

(b) What would be the highest CO concentration in the residential zone (i.e., the worst-case scenario)?

2. An industrial city is to be developed in a valley. The entire city area would be a rectangular block (5 km × 30 km). It is to have three zones of the same size (5 km × 10 km each), one for residential, one for light-industrial, and one for heavy industrial. Use the area-source box model and the following simplified assumptions to answer the questions below.

- The wind blows continuously and parallel to the long-side of the city block at 2 m/s and brings SO_2 from upstream at a concentration of 0.1 milligram/m^3.

- The SO_2 emission rates are 0.2 and 2 g/km^2/d from the light and the heavy industrial zones, respectively. The mixing height = 500 m (constant)

(a) The sequences of these three zones which would provide the best and the worst ambient SO_2 concentrations in the residential zone.

(b) The lowest potential ambient SO_2 concentration in the residential zone.

(c) The highest potential ambient SO_2 concentration in the residential zone.

3. An industrial boiler with a stack height of 80 m is emitting 4 g/m^3 SO_2 with a volumetric flow rate of 25 m^3/s on a sunny summer day (diameter of the stack = 2 m). The wind speed at the top of the stack is 4 m/s. The temperature of the exit gas is 120 °C. The ambient temperature is 30 °C and pressure is 750 mm-Hg.

(a) Determine the effective stack height

(b) Determine the maximum ground level concentration at one kilometer downstream of the stack

(c) Determine the distance where the maximum ground level concentration occur

(d) Determine the concentration at location found in part (c).

4. A plant is emitting 300 g/s of particulates and the effective stack height is 100 m. The wind speed is 3 m/s.

   (a) What would be the locations of the maximum ground concentrations when the stability classes are C and D, respectively?

   (b) What would the ratio of these two maximum ground level concentrations?

   (c) What is the maximum ground concentration for air stability class C?

5. The SO_2 emission rate from a stack is 800 g/s (the effective stack height = 40 m). On a specific day, the Pasquill stability class of the atmosphere is D and the wind speed = 3 m/s.

   (a) What would be the maximum ground-level SO_2 concentration at 1 km downwind from the stack?

   (b) What would be the maximum ground-level SO_2 concentration from this discharge?

   (c) How far downwind will the maximum concentration in part (b) located?

Chapter 6

Control of Particulate Emissions

Particulate matter (PM) is one of the six criteria pollutants. Reducing particulate emissions from stationary source is vital to achieve good ambient air quality. There are many types of control devices available, and removal mechanisms of these devices vary. Characteristics of particulates and of the gas stream will have significant impacts on their removal.

This chapter starts with an introduction on sources of particulate emissions, particulate removal and types of commonly-used removal devices, and characteristics of gas stream and particulate that are relevant to particulate removal (Section 6.1). Section 6.2 describes particle size, particle size distribution, particulate Reynolds numbers and detailed discussion on particulate removal mechanisms. Principles, commonly-used systems and their components, important design considerations, and design calculations related to gravity settling chambers are presented in Section 6.3. Those of cyclones, electronic precipitators, fabric filters, and wet scrubbers are presented in Sections 6.4, 6.5, 6.6, and 6.7, respectively. Section 6.8 summarizes important considerations in selection of proper particulate removal devices.

6.1 Introduction

Particulate matter (PM) is a general term that describes solid particles and liquid droplets present in the atmosphere and gas streams. An obvious effect of PM pollution is reduction of visibility. Emission from a stack or a tail pipe is visible mainly because of the PM it carries. Elevated PM concentrations in the ambient air also pose threats to human health, ecosystem, and the environment.

6.1.1 Sources of particulate emissions
Particles are being emitted to the atmosphere from natural processes and human activities. Some particles are formed in the atmosphere. Solid particles can be generated from mechanical action (e.g., grinding), chemical reactions (e.g., metal oxides), and incomplete combustion (e.g., carbon black). Liquid particles can be generated from chemical reactions and condensation.

Primary particles are those released directly from sources to the atmosphere. They include wind dusts, ocean sprays, volcano ash, and carbon black from combustion sources. *Secondary particles* are those formed in the atmosphere from chemical reactions involving primary gaseous pollutants. Examples of secondary particles are ammonium sulfate ($(NH_4)_2SO_4$) and ammonium nitrate (NH_4NO_3), formed from gaseous emissions from various combustion sources. When compared to primary particles, they are much smaller in size and represent a large portion of $PM_{2.5}$ in our ambient air.

Tables 6.1 and 6.2 tabulate the PM_{10} and $PM_{2.5}$ emissions from main sources categories in the U.S. in 2017. The total emissions are 18.1 and 5.3 million tons for PM_{10} and $PM_{2.5}$, respectively. Contributions to PM_{10} emissions are 86.6, 4.8, 6.1, and 2.5% from miscellaneous, stationary fuel combustion, industrial and other processes, and transportation, respectively. Contributions to $PM_{2.5}$ emissions are 69.3, 14.0, 11.3, and 5.4% from miscellaneous, stationary fuel combustion, industrial and other processes, and transportation, respectively. Miscellaneous source category (e.g., wildfires, road dusts, ocean sprays) accounts for majority of both PM_{10} and $PM_{2.5}$ emissions.

Figures 6.1 and 6.2 depict the trends of PM_{10} and $PM_{2.5}$ annual emissions in the U.S. since 1990. As shown, the emissions of both have been decreasing, and most of the reductions come from the major source category, the miscellaneous sector.

Table 6.1 - PM$_{10}$ emission rates and source categories in the U.S. in 2017

Source Category	(1,000 tons)	(%)
Stationary fuel combustion	**867**	**4.8**
Electric utility	234	1.3
Industral	286	1.6
Others	348	1.9
Industrial and other processes	**1,112**	**6.1**
Chemical & allied product manufacturing	19	0.1
Metals processing	58	0.3
Petroleum & related industries	33	0.2
Other industrial processes	677	3.7
Solvent utilization	4	0.0
Storage & transport	43	0.2
Waste disposal & recycling	277	1.5
Transportation	**446**	**2.5**
Highway vehicles	261	1.4
Off-Highway	185	1.0
Miscellaneous	**15,727**	**86.6**
Wildfires	1,046	5.8
Others	14,681	80.9
Total	18,152	100.0

Table 6.2 - PM$_{2.5}$ emission rates and source categories in the U.S. in 2017

Source Category	(1,000 tons)	(%)
Stationary fuel combustion	**749**	**14.0**
Electric utility	182	3.4
Industral	224	4.2
Others	343	6.4
Industrial and other processes	**603**	**11.3**
Chemical & allied product manufacturing	14	0.3
Metals processing	44	0.8
Petroleum & related industries	29	0.6
Other industrial processes	265	5.0
Solvent utilization	4	0.1
Storage & transport	17	0.3
Waste disposal & recycling	230	4.3
Transportation	**290**	**5.4**
Highway vehicles	116	2.2
Off-Highway	174	3.3
Miscellaneous	**3,700**	**69.3**
Wildfires	886	16.6
Others	2,814	52.7
Total	**5,343**	**100.0**

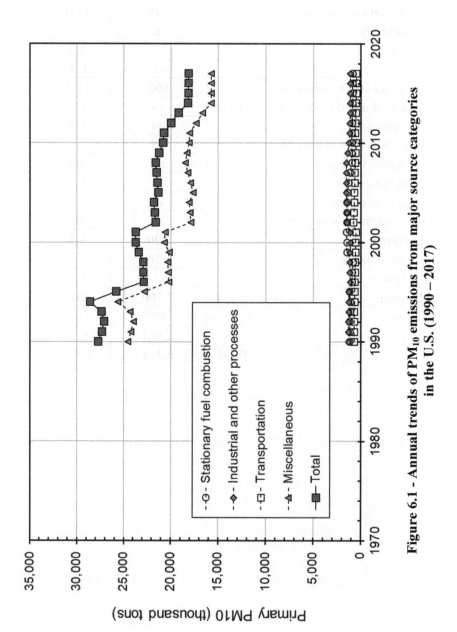

Figure 6.1 - Annual trends of PM$_{10}$ emissions from major source categories in the U.S. (1990 – 2017)

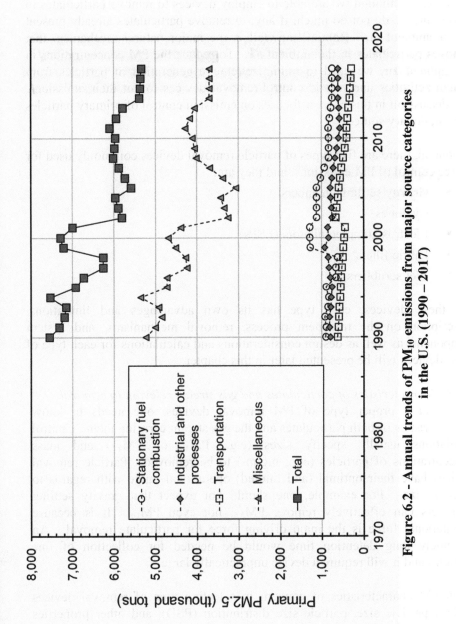

Figure 6.2 - Annual trends of PM$_{10}$ emissions from major source categories in the U.S. (1990 – 2017)

6.1.2 Particulate removal and commonly-used removal devices

Not much we can do with regards to reducing PM emissions from natural processes. Although we are able to employ devices to remove particulates in indoor air, we do not do much, if any, to remove particulates already present in our ambient air. Rainfall/snowfall is the major natural mechanism that removes particulates in the ambient air. To reduce the PM concentrations in the ambient air, we need to minimize/remove generation of particles from human activities and to add control/removal devices to cut their emissions. The discussion in this chapter focuses on emission control of primary particles from stationary sources.

In general, there are five types of particle removal devices commonly used for source control of PM emissions; and they are:

- Gravity settling chambers
- Cyclones
- Electrostatic precipitators (ESPs)
- Fabric filters
- Wet scrubbers

For these devices, each type has its own advantages and limitations. Description on the treatment process, removal mechanisms, and system components as well as design considerations and calculations for each type of these devices will be presented later in this chapter.

6.1.3 Characteristics of particulates and gas stream relevant to removal

To select a proper type of PM removal devices, one needs to know characteristics of both particulates and the gas stream carrying them. Control regulations usually specify sizes (e.g., PM_{10} or $PM_{2.5}$) and mass concentrations of particles (e.g., mg/m^3) to be removed. Particle removal devices have their optimal (and limited) operational range with regards to particle size. For example, one should not expect that gravity settling chambers can effectively remove $PM_{2.5}$, not even PM_{10}. It is because gravitational force is the main driving force for particulate removal. An extremely long detention time would be needed for collection of fine particles and it will require a device unpractically large.

Main PM characteristics of concern in selecting type of removal devices include particle size, particle size distribution (PSD), and other properties (e.g., shape, density, sticky, oily, acidic/caustic, etc.). For example, sticky particles may not be good for fabric filters because they may bind to fabric filters and also present a fire hazard.

116

A gas stream, carrying particulates of concern, needs to move through the PM removal device. The flow rate into the control device and its resultant velocity have significant impacts on sizes of particles that can be removed at a desirable/design efficiency. The gas stream flow rate is positively related to the size of the device as well as head-loss (pressure drop) across the device. A higher rate would require a larger device, a larger blower and incur a higher energy cost.

Temperature of the influent gas stream is also an important parameter. For example, fabric materials used in some filters may not be able to stand elevated temperatures. High humidity in the influent gas stream may present a challenge to some types of control devices, but it is totally acceptable for wet scrubbers. In addition, presence of some trace constituents may have significant adverse impacts. For example, presence of hydrogen sulfide (H_2S) would lower the dew point of the gas stream and acidic condensate may form when the gas stream is cooled in gravity chambers [Note: the temperature of a gas stream will drop upon expansion].

6.2 Particle Size, Size Distribution, and Removal Mechanisms

6.2.1 Particle size
Particles in air can be of different shapes such as a perfect solid sphere, hollow sphere, irregular solid, flake, fiber, condensation floc, and aggregate (Figure 6.3).

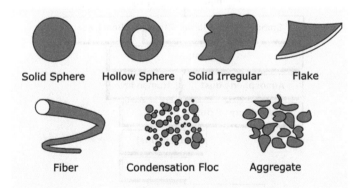

Figure 6.3 - Shapes of typical airborne particles (EPA, 2012)

For practical applications, we often assume that particles are spherical. An equivalent particle diameter ($D_{p,equivalent}$) is often used, assuming that a spherical particle of this diameter would have the same behavior as the particle in question. For example, if the volume is the one of the concern, its equivalent diameter can be found by using its volume ($V_{particle}$) as:

$$D_{p,equivalent} = \sqrt[3]{(6/\pi) \times V_{particle}} \qquad (6.1)$$

Equation 6.1 is based on the fact that the volume of a perfect sphere is equal to $(\pi/6)D_p^3$.

For air pollution control, ranges of particle sizes that need to be dealt with are broad, typically in the range of hundredths to a few hundreds micrometers. Micrometer (μm, or *micron (μ)*) is one millionth of a meter and it is a common unit for diameters of small particles. For example, PM_{10} represents particles with aerodynamic diameters ≤10 μ (aerodynamic diameter will be defined later) and they are considered "respirable". Figure 6.4 shows sizes of typical airborne particles compared to some common materials. Figure 6.5 illustrates size ranges of particles generated by several particle formation mechanisms. Physical attraction and ash burnout are mechanisms for formation of primary particles while heterogeneous nucleation and homogeneous condensation are for secondary particles.

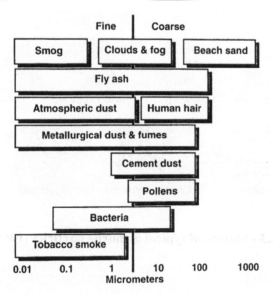

Figure 6.4 - Sizes of airborne particles (EPA, 2013)

118

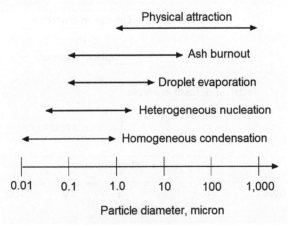

Figure 6.5 - Size ranges of particles from several formation mechanisms (Modified from EPA, 2012)

Since the surface area and volume of a sphere are πD_p^2 and $\pi D_p^3/6$, respectively; the surface and volume of a 10-μ particle are 100 and 1,000 times of those of a 1-μ particle, respectively. The size of a particle will affect its behavior in ambient air as well as in removal devices. Particles with diameters between 1 and 10 μm are especially important in air pollution control. Very fine particles (0.01 - 0.1 μm) can be generated in processes such as combustion. However, they often rapidly agglomerate into larger particles (>0.1 μm). Particles in the range of 0.1 to 1.0 μm can be a significant portion of particulate emissions from many sources and they are relatively hard to collect (EPA, 2003). Sometimes, the particles are classified into categories of super-coarse, coarse, fine, and superfine for aerodynamic diameter >10, between >2.5 and ≤ 10, >0.1 and ≤2.5, and ≤ 0.1μ, respectively.

There are several types of particle size measurement devices, including microscope, optical counter, electrical aerosol analyzer, Bahco counter, and inertial impactor (Figure 6.6). Detailed descriptions of these instruments can be found in EPA (2012).

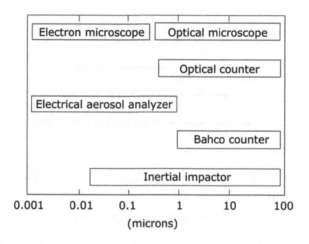

**Figure 6.6 - Ranges of particle size measuring devices
(Modified from EPA, 2012)**

6.2.2 Particle size distribution (PSD)

A gas stream usually carrying particles of different sizes with different concentrations, one of the simplest ways to describe its particle size distribution (PSD) is a histogram (Figure 6.7). Frequency of occurrence in a histogram can be number, surface area, volume, or mass of the particles. In Figure 6.7, the *mode* is the particle size that occurs most frequently. If the frequency refers to number of particles, the *median* particle size divides the frequency distribution into two halves (50% of the particles with their diameters larger and the other 50% with their diameters smaller than this diameter) and it corresponds to a cumulative fraction of 50%. The *mean* diameter is the arithmetic average of the diameters of all particles. The value of the mean (i.e., the arithmetic average) is more sensitive, than the median, to the quantities of the particles in the extremely lower and/or upper ends of the distribution. They are often considered "outliers".

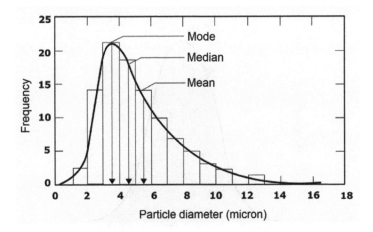

**Figure 6.7 - Particle size distribution
(Modified from EPA, 2012)**

A simplified saying - if the curve shown in Figure 6.7 is a symmetrical bell curve, then the PSD has a *normal (or Gaussian) distribution.* For many stationary and mobile sources, the observed PSD of exhaust gas streams are log-normally distributed. It means that when the frequency of occurrence is plotted against logarithm of the particle diameter, a symmetrical bell curve would be generated (Figure 6.8). The frequency on the vertical scale can be percentage of mass divided by the difference in the logarithms of the particle sizes defining a specific interval, which is $[\log(D_{p,max}) - \log(D_{p,min})]$, or $\Delta\log$ D_p. A plot of log-normally distributed PSD on a log-probability paper will be a straight line. Similarly, a plot of normally-distributed PSD on a regular probability paper will also be a straight line. The data can be plotted in terms of either the cumulative percentage of particles smaller or larger than a specific size. In this book the phrase "the percentage smaller" would be used.

121

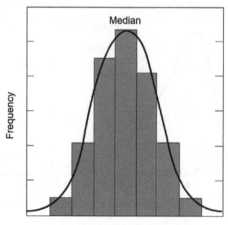

Figure 6.8 - Log-normal particle size distribution
(Modified from EPA, 2012)

A straight line on a log-probability paper for a log-normally distributed particle mass can tell the mean diameter and standard deviation of this set of particles. The mean diameter is the particle size corresponds to the 50% probability. For a log-normally distributed particle mass, its *standard deviation* (σ) and *geometric standard deviation* (σ_g) can be found from its geometric mean diameter (D_{50}) and the particle size at the 15.87% probability ($D_{15.87}$) or that at the 84.13% probability ($D_{84.17}$) as:

$$\sigma = \ln(\sigma_g) = \ln\left(\frac{D_{84.13}}{D_{50}}\right) = \ln\left(\frac{D_{50}}{D_{15.87}}\right) \qquad (6.2)$$

The standard deviation (σ) should be greater than or equal to the geometric standard deviation (σ_g); and σ_g should be ≥ 1. For the special case of $\sigma = 0$ (or $\sigma_g = 1$), all the particles in the entire particle set have the same size.

Example 6.1   A log-normally distributed particle mass

A log-normally distributed particle mass has a geometric mean diameter of 15 microns and a geometric standard deviation of 5 microns. Determine the particle diameter that 15.87% of the entire particle mass having diameters smaller than it?

Solution:

$$\ln(\sigma_g) = \ln(5) = \ln\left(\frac{15}{D_{15.87}}\right)$$

$$D_{15.87} = \underline{3\mu}$$

Discussion: Using a similar approach, you should be able to find that 84.13% of the entire particle mass having diameters $\leq 75\mu$.

Example 6.2 Analysis of particle size distribution

Determine if sizes of particles in the table below are log-normally distributed. If so, find the geometric mean diameter and the geometric standard deviation.

Particle size (μ)	Mass (mg)
<0.5	0.10
0.5 to 1.0	0.15
1.0 to 2.0	0.20
2.0 to 4.0	0.20
4.0 to 6.0	0.15
6.0 to 8.0	0.10
>8.0	0.10

Solution:

(a) Calculate the fraction and the cumulative fraction of each size range as:

Particle size (μ)	Mass (mg)	Fraction of mass in this range	Cumulative fraction
<0.5	0.10	0.10	0.10
0.5 to 1.0	0.15	0.15	0.25
1.0 to 2.0	0.20	0.20	0.45
2.0 to 4.0	0.20	0.20	0.65
4.0 to 6.0	0.15	0.15	0.80
6.0 to 8.0	0.10	0.10	0.90
>8.0	0.10	0.10	1.00
Total	1.00		

(b) Plot $D_{p,max}$ of each size range versus its corresponding cumulative fraction smaller than it on a log-probability paper:

123

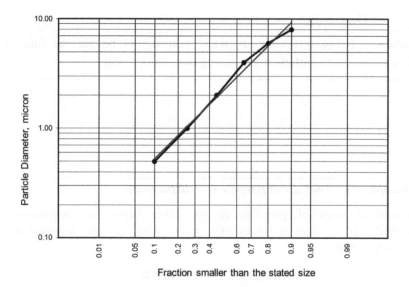

Fraction smaller than the stated size

(c) As shown, a straight line fits the data relatively well. The sizes of this set of particles appear to be log-normally distributed.

(d) From the plot, $D_{50} = 2.3\mu$ and $D_{84.12} = 7.0\mu$:

$\sigma_g = (7.0/2.3) = \underline{3.05\mu}$

Figure 6.9 illustrates the results of a study on PM concentrations in ambient air and the measurements were made in traffic. The x-axis is the particle diameter (in μ) on a logrithmetic scale and the y-axis is $\Delta V/\Delta logD_p$, which is the volume of the particles (in μm^3) per cm^3 of air having diameters in the size range from $logD_p$ to $log(D_p + \Delta D_p)$. The dotted vertical line at $D_p = 1\mu$ (arbitrarily chosen) divides the particles into coarse and fine particles, based on the particle size. The particles were also grouped into three modes: nuclei, accumulation, and coarse, based on their formation mechanisms. Particles of all three modes are geometrically distributed. The coarse-mode particles were mechanically generated (e.g., from wind erosion of crustal materials). For the particles of this mode, the diameter of geometric-mean by volume (DGV) is 4.9μ and the geometric standard deviation (σ_g) is 1.87μ. The particles smaller than 1μ were generated in combustion or formed from gas vapors through accumulation or condensation. The DGV and σ_g of the nuclei-mode particles are 0.018 and 1.6μ, respectively. The portion of the fine particles with ~0.1μ < D_p < ~1μ were formed from accumulation (DGV = 0.21μ and $\sigma_g = 1.8\mu$).

124

Nuclei-mode particles might grow from coagulation/flocculation into large sizes to be in the accumulation mode (Wilson *et al.*, 2002).

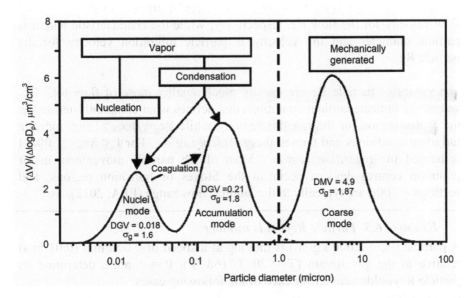

Figure 6.9 - Size distribution and formation mode of PM in ambient air (modified from EPA, 2012)

Densities of a specific type of particles are often assumed to be the same. If this is the case, the volume fraction for a size range should be the same as the corresponding weight/mass fraction. For particulate control, we are more interested in removal of particulate mass, rather than the number of particles. However, we need to have an idea on the sizes of particles need to be removed. Most, if not all, of the control devices are more effective in removing larger particles. Effectiveness in removing particles of specific size ranges is one of the major considerations in selecting control devices for a pollution control project. Discussions hereafter will focus more on removal of mass, instead of number, of particles.

6.2.3 Reynolds number for particulate travel
In Chapter 2, we talked about Reynolds number (Re) for a flowing fluid, which sometimes termed (*flow*) *Reynolds Number*. To characterize conditions for particles flowing through a fluid or with a moving fluid, (*particle*) *Reynolds Number* (Re$_p$) is defined as:

$$Re_p = \frac{D_p \times u_p \times \rho_f}{\mu_f} \qquad (6.3)$$

where D_p = particle diameter, u_p = particle velocity relative to the fluid, ρ_f = density of the fluid, and μ_f = viscosity of the fluid. It should be noted that the characteristic length is duct diameter (for a circular pipe) and velocity is fluid flow velocity for the flow Re, respectively; while the characteristic length is particle diameter and the velocity is particle migration velocity for the particle Re.

The values of particle Re are usually much smaller those of flow Re. Re_p values < 1 indicate laminar conditions that defines what commonly-termed as the *Stokes* region (or the *laminar region*); while Re_p values >1,000 indicate turbulent conditions and termed the *turbulent region*. For $1 < Re_p < 1,000$, t is termed the *transition region*. Most of the particle movements in air pollution control devices occur in the Stokes or transition regions; and particles < 100μ are typically in the laminar flow range (EPA, 2012).

Example 6.3 Particle Reynolds number

A particle (ρ_p = 2,000 kg/m³) is moving at a speed of 0.6 m/min (0.01 m/s) relative to the gas stream (T = 20 °C (68 °F), P = 1 atm), determine its particle Reynolds number for each of the following cases:

(a) $D_p = 10\mu$
(b) $D_p = 100\mu$
(c) $D_p = 10\mu$, but $\rho_p = 4,000$ kg/m³
(d) $D_p = 10\mu$, but T = 200 °C and P = 1 atm.

Solution:

(a) Assuming the gas stream has the same density and viscosity as air, then

$$Re_p = \frac{(10 \times 10^{-6}m)(0.01\frac{m}{s})(1.21\frac{kg}{m^3})}{(1.81 \times 10^{-5}\frac{kg}{m-s})} = 6.7 \times 10^{-3}$$

(a) Re_p is proportional to the particle diameter, so Re_p should be 10 times larger, $\underline{6.7 \times 10^{-2}}$.

(b) Re_p is independent of the density of the particle, so $Re_p = \underline{6.7 \times 10^{-3}}$.

(c) Re_p is proportional to the density of the fluid, but inversely proportional to the viscosity of the fluid. Using the values of the air density and viscosity from Examples 2.2 & 2.3:

$$\frac{Re_{p,2}}{Re_{p,1}} = \left(\frac{Re_{p,473K}}{6.7\times10^{-3}}\right) = \left(\frac{\rho_{f,2}}{\rho_{f,1}}\right)\left(\frac{\mu_{f,1}}{\mu_{f,2}}\right) = \left(\frac{0.75}{1.21}\right)\left(\frac{1.81\times10^{-5}}{4.73\times10^{-5}}\right)$$

$Re_{p,473K} = \underline{4.15 \times 10^{-3}}$

Discussion: Re_p of the particle decreased as the temperature increases because of the decrease in the air density as well as the increase in the air viscosity.

6.2.4 Particle collection/removal mechanisms

Five types of particulate removal devices are covered in this book: gravity settling chambers, cyclones, fabric filters, electrostatic precipitators (ESPs), and wet scrubbers. For each type, particles are collected/removed by one or more mechanisms. The removal mechanisms of one type of devices may be similar and/or different from those of the others. In addition, one mechanism may play a more significant role than the others for a given type of devices. This section discusses several important removal mechanisms individually. The particulate removal mechanisms involved in each type of devices will be mentioned later. The main particulate removal mechanisms include gravity settling, centrifugation, Brownian motion/diffusion, electrostatic attraction, and inertial impaction.

Gravity Settling. For a discrete particle to settle down in a fluid, it would be subjected to three forces: gravitational force (F_G), buoyancy force (F_B), and drag force (F_D).

For a spherical particle, F_G can be expressed as

$$F_G = m_p g = \left(\rho_p \times V_p\right)g = \left(\rho_p \times \frac{\pi D_p^3}{6}\right) g = \frac{\pi D_p^3 g}{6} \times \rho_p \quad (6.4)$$

where m_p, ρ_p, and V_p are the mass, density, and volume of the particle, respectively; and g is the gravitational constant.

Against the particle's downward movement, the buoyancy force would be

$$F_B = \frac{\pi D_p^3 g}{6} \times \rho_a \quad (6.5)$$

where ρ_a is the density of the air.

The drag force (F_D) can be considered as the resistant force due to shearing stress as an object moving through a fluid. In general, it can be expressed as

$$F_D = C_D \times \rho_f \times \left(\frac{v_p^2}{2}\right) \times A \quad (6.6)$$

where C_D is the drag coefficient, ρ_f = density of the fluid, and A = characteristic frontal area of the object (= $\pi D^2/4$ for a sphere). Equations have been derived to relate Re_p with C_D. Most of them in literature are the same for the laminar region and similar for the turbulent region, while they appear to be different for the transition region. As shown in EPA (2012),

$$C_D = {24}/{Re_p} \qquad \text{for } Re_p < 1 \qquad\qquad (6.7)$$

$$C_D = {18.5}/{Re_p{}^{0.6}} \qquad \text{for } 1 < Re_p < 1{,}000 \qquad (6.8)$$

$$C_D = 0.44 \qquad \text{for } Re_p > 1{,}000 \qquad\qquad (6.9)$$

Under the laminar flow region ($Re_p < 1$), the drag force (F_D) can be expressed as (derived by Sir George Stokes in 1851):

$$F_D = 3\pi\mu D_p u \qquad (6.10)$$

where u is the particle migration velocity.

When the system is in equilibrium, the sum of all the external forces (i.e., F_G, F_B, and F_D) is equal to zero. Thus,

$$\Sigma F = F_B + F_G + F_D = \frac{\pi D_p^3 g}{6} \times \rho_a - \frac{\pi D_p^3 g}{6} \times \rho_p + 3\pi\mu D_p u = 0 \quad (6.11)$$

The "minus" sign in Equation 6.11 accounts for the fact that the gravitational force is in the opposite direction of the buoyancy and the drag forces.

In equilibrium, the particle will stand still or settle down at a constant velocity, which is often termed the *terminal velocity* (u_t):

$$u_t = \frac{g \times (\rho_p - \rho_a) \times D_p{}^2}{18\mu} \approx \frac{g \times \rho_p \times D_p^2}{18\mu} \qquad (6.12)$$

The equation is often referred to as the *Stokes Law* or the *Stokes equation* (note: $\rho_p \gg \rho_a$).

If a particle has $D_p > 3\mu$, the gas would appear to be "continuous" around the particle. It means that this particle will not be affected by collisions with individual gas molecules because this type of collision occurs frequently on all sides of this particle. On the other hand, if a particle having $D_p < 3\mu$, the fluid would appear to be "discontinuous" and this often occurs for small particles in the laminar flow region ($Re_p < 1$). In this case, the collisions with gas molecules would cause the particle to move in a direction related to the resultant force acting on the particle. The particle will "*slip*" between the gas molecules. For gravitational settling, the particle will move faster than what Eq. 6.12 would predict (EPA, 2012). To correct for this, *Cunningham slip correction factor* (C_c) is often used to reduce the drag coefficient in the laminar flow region as:

$$C_D = \left[24/Re_p \right] / C_c \qquad (6.13)$$

The drag force in Eq. 6.10 will also be reduced by a factor of C_c as:

$$F_D = \frac{3\pi\mu D_p u}{C_c} \qquad (6.14)$$

Consequently, the terminal velocity in Eq. 6.12 can be modified to:

$$u_t = C_c \left[\frac{g \times (\rho_p - \rho_a) \times D_p{}^2}{18\mu} \right] \approx C_c \left[\frac{g \times \rho_p \times D_p^2}{18\mu} \right] \qquad (6.15)$$

It should be noted the Stokes Law (Eq. 6.12) will under-estimate the terminal velocities of sub-micron particles (as shown in Eq. 6.15), but over-estimate them for particles having diameters of tens of microns or larger.

An empirical equation can be used to estimate values of Cunningham slip correction factor (EPA, 2012):

$$C_c = 1 + \frac{6.21 \times 10^{-4} T}{D_p} \qquad (6.16)$$

where T = absolute temperature (in K) and D_p = particle diameter (in μ).

As shown in Eq. 6.16, values of C_c are always larger than unity. They increase with T, but decrease with D_p. At 293K (20 °C), the values of C_c are 2.8, 1.18, and 1.06 for $D_p = 0.1$, 1 and 3μ, respectively. The Cunningham slip correction factor becomes increasingly significant when the particle diameter becomes <3 μ.

Example 6.4 Terminal settling velocity of a particle

A spherical particle ($D_p = 2\mu$, $\rho_p = 2,000$ kg/m^3) is settling in a still air ($T = 20$ °C (68 °F), $P = 1$ atm) by gravity.

(a) Estimate the value of the Cunningham slip correction factor.
(b) Estimate the terminal settling velocity (u_t) of this particle.
(c) What would be the u_t value, if $D_p = 100\mu$?
(d) What would be the u_t value, if $D_p = 2\mu$ and 4,000 kg/m^3?

Solution:

(a) From Eq. 6.16,

$$C_c = 1 + \frac{6.21 \times 10^{-4}T}{D_p} = 1 + \frac{(6.21 \times 10^{-4})(293)}{2} = 1.09$$

(b) From Eq. 6.15,

$$u_t = (1.09)\left[\frac{(9.81)(2,000 - 1.21)(2 \times 10^{-6})^2}{(18)(1.81 \times 10^{-5})}\right] = 2.41 \times 10^{-4} \ m/s$$

(c) For a 100μ particle, its u_t value would be 0.6 m/s, 2,500 times of the 2μ particle, because u_t is proportional to D_p^2.

(d) The u_t value will be doubled, because u_t is proportional to ($\rho_p - \rho_a$), or proportional to ρ_p since $\rho_a \ll \rho_p$.

Discussion:

1. Eq. 6.16 is an empirical equation; values need to be used with the specified units.

2. Settling velocities of particles of micron sizes are small, as shown in part (a), but it becomes much larger for a 100μ particle.

To evaluate the motion of a particle in air or a control device, the *aerodynamic diameter* (D_a) is often used as:

$$D_a = D_p\sqrt{\rho_p C_c} \qquad (6.18)$$

For common uses, D_p is the particle diameter in μ and ρ_p is the density of particle in g/cm^3. With these, the aerodynamic diameter will have a unit of μ_a, which stands for "micron, aerodynamic". Particles of the same aerodynamic diameter will behave identically in several types of particulate removal devices. The aerodynamic diameter can be defined as the diameter of a spherical particle with a density of 1,000 kg/m^3 (1 g/cm^3), but with the same settling velocity as that of the particle in question (Wilson et al., 2002). It should be noted that the diameters used for definitions of $PM_{2.5}$ and PM_{10} are actually aerodynamic diameters, not the physical ones.

Example 6.5 Aerodynamic diameter

Determine the aerodynamic diameter of a particle (a) $D_p = 20\mu$ and $\rho_P = 1,000$ kg/m^3 and (b) $D_p = 10\mu$ & $\rho_P = 4,000$ kg/m^3, assuming $C_c = 1$.

Solution:

(a) From Eq. 6.17,

$$D_a = D_p\sqrt{\rho_p C_c} = (20)\sqrt{(1)(1)} = 20\ \mu_a$$

(b) Also from Eq. 6.17,

$$D_a = (10)\sqrt{(1)(4)} = 20\ \mu_a$$

Discussion:

1. Density values used in Eq. 6.17 should be in g/cm^3.

2. These two particles have the same aerodynamic diameter, even one diameter is twice as large as the other.

Centrifugation. A particle within a gas stream travelling in a circular motion is experiencing a centrifugal force (F_c), which can be expressed as

$$F_c = m\left(\frac{v^2}{r}\right) \qquad (6.17)$$

where v = tangential velocity of the gas stream and r = radial position of the particle. With $r = 0.5$ m and $v = 10$ m/s, the acceleration (v^2/r) would be 200 m^2/s, which is 20.4 (= 200/9.81) times of the gravitational acceleration.

For particles in the laminar flow region and assuming a particle is moving at the same speed of the gas stream (v_{gas}) in a circular motion, the tangential velocity (u_p) of this particle can be determined by:

$$u_p = C_c\left[\frac{\rho_p D_p^2}{18\mu}\right]\left(\frac{v_{gas}^2}{r}\right) \qquad (6.19)$$

Eq. (6.15) and Eq. (6.19) are similar, except the acceleration terms are different (i.e., g versus v_{gas}^2/r).

Example 6.6 Tangential velocity of a particle in a cyclone

Referring to the particle in Example 6.4, if it enters a cyclone (diameter = 1 m) with a gas stream at a velocity of 10 m/s. Determine the tangential velocity of this particle.

Solution:

Using Eq. 6.19,

$$u_p = (1.09) \left[\frac{(2,000)(2 \times 10^{-6})^2}{(18)(1.81 \times 10^{-5})} \right] \left(\frac{10^2}{0.5} \right) = 4.91 \times 10^{-3} \; m/s$$

Discussion: The particle's tangential velocity caused by the centrifugal force is 4.91×10^{-3} m/s, which is 20.4 times of the terminal velocity due to the gravitational force (2.41×10^{-4} m/s).

Electrostatic attraction. Electrostatic attraction is the main mechanism responsible for particulate removals by ESPs. For the purpose of discussion here, let us envision a simplified wire-and-plate ESP as having two parallel plates with several electrodes sitting in the middle plane between these two plates. A strong electrical field is established between the discharge electrodes and the plates are electrically-grounded.

Electrical discharges from a discharge electrode are termed *corona discharge*. With the corona discharge, released electrons are accelerated by the strong electric field and move toward the electrically-grounded collection plates. On their way, some of these electrons are captured by gas molecules and these gas molecules become negatively charged; they are often called *gas ions*. On their way to the collection plates, these gas ions may be intercepted/captured by larger particles ($> \sim 1\mu$) because these particles locally disrupt the electrical field. The negatively-charged particles will then start moving toward the collection plates. They may quickly reach the *saturation* charge *(or the equilibrium charge)"* which is strong enough to deflect additional gas ions. The magnitude of the saturation charge is a function of the particle surface area. Large particles would accumulate higher electrical charges because of their larger surface areas and, consequently, they will be more strongly affected by the applied electrical field. This is *field charging* (or *contact charging*) and it is the dominant charging mechanism for particles having $D_p > 2\mu$ (EPA, 2003; EPA, 2012).

Field charging becomes less important as the particle size decreases because they do not have sufficient mass/size to disrupt the electrical field. For particles having D_p <0.4μ, the diffusional charging becomes more important. *Diffusional charging* (or *ion charging*) results from collisions caused by the random Brownian motion of both gas ions and particles, not due to the electrical field (EPA, 2003; EPA, 2012).

The strength of an electrical field (E) can be defined as the applied voltage (V) over the distance (*x*) as:

$$E = \frac{\partial V}{\partial x} \qquad (6.20)$$

A typical unit for E is Volt/m (or V/m).

The saturation charge (q_{sat}) of a spherical particle could be expressed as:

$$q_{sat} = 3\pi \left[\frac{\kappa}{\kappa+2}\right] \varepsilon_0 D_p^2 E_0 \qquad (6.21)$$

where ε_0 = permittivity of air and E_0 = strength of the electric field.

In electromagnetism, *permittivity* is a measure of a material's ability to transmit an electrical field in a particular medium. Opposite to the literal meaning of "permittivity", a charge will actually yield more electric flux in a medium with a lower permittivity. The lowest permittivity is that of a vacuum (free space) and it has a value of 8.85 × 10^{-12} Coulomb/V-m. *Relative permittivity* (or *dielectric constant*), κ, is usually used for a medium other than the vacuum, which is the ratio of the permittivity of the medium (ε) to that of the vacuum (ε_0). The dielectric constant of air is 1.0006, which is very close to that of the vacuum, which is 1.0 [Note: the dielectric constant of water @ 20 °C (68 °F) = 80.4]. Typical values of dielectric constant for solid particles are from 4 to 8.

Coulomb is the SI unit for electric charge, and one Coulomb (C) is equal to the quantity of electricity conveyed by a current of one ampere in one second (= 6.24 × 10^{18} electrons). Consequently, the charge of one electron is 1.602 × 10^{-19} C.

Example 6.7 Saturation charge of a particle in an electrical field

Referring to the particle in Example 6.4, it has reached its saturation charge in an ESP at a location where the field strength is 320 kV/m. Assuming the dielectric constant of this 2μ particle is 5, how many electronic charges on the surface of this particle at saturation?

Solution:

Using Eq. 6.21,

$$q_{sat} = 3\pi \left[\frac{5}{5+2}\right] \left(8.85 \times 10^{-12} \frac{C}{V \cdot m}\right) (2 \times 10^{-6} \ m)^2 \left(320{,}000 \frac{V}{m}\right)$$

$$= (7.73 \times 10^{-17} C) = (7.73 \times 10^{-17} C) \times \left(\frac{6.24 \times 10^{18}}{C}\right) = 476 \ electrons$$

Discussion: Since the saturation charge is proportional to D_p^2, the number of electrons on a 1μ particle will be 119 (i.e., one quarter of 476).

The electric force on a particle (F_E) is the product of its surface charge and the local electric field strength (E_p) as:

$$F_E = q \times E_p = \left\{ 3\pi \left[\frac{\kappa}{\kappa+2}\right] \varepsilon_0 D_p^2 E_0 \right\} \times E_p \tag{6.22}$$

It is a common practice to use an average E to replace both the field strength at the time when the particle is being charged (E_0) and that at the location the particle currently locates (E_p), Eq. 6.22 can then be simplified to:

$$F_E = q \times E_p = 3\pi \left[\frac{\kappa}{\kappa+2}\right] \varepsilon_0 D_p^2 E^2 \tag{6.23}$$

If the Stokes drag forces (Eq. 6.14) is the only resistance to the particle movement induced by the electrostatic force, then the terminal velocity (u_t) will occur when the electrostatic force and the drag force are equal. Then,

$$F_E = E_D = 3\pi \left[\frac{\kappa}{\kappa+2}\right] \varepsilon_0 D_p^2 E^2 = \frac{3\pi \mu D_p u}{C_c} \tag{6.24}$$

The terminal velocity can be readily derived by rearranging Eq. (6.24):

$$u_t = \frac{\left[\frac{\kappa}{\kappa+2}\right] \varepsilon_0 D_p E^2}{\mu} = w \tag{6.25}$$

Drift velocity, instead of terminal velocity, is commonly used in the ESP literature, and the symbol w is commonly used for the drift velocity.

Example 6.8 Drift velocity of a particle in an electrical field
Referring to the particle in Example 6.7, find its drift velocity.

Solution:

Using Eq. 6.25,

$$w = \left\{ \left[\frac{5}{5+2}\right] \left(8.85 \times 10^{-12} \frac{C}{V \cdot m}\right) (2 \times 10^{-6} m) \left(320,000 \frac{V}{m}\right)^2 \right\}$$

$$\div \left(1.81 \times 10^{-5} \frac{N \cdot s}{m^2}\right) = 7.15 \times 10^{-2} m/s$$

Discussion:

1. $1 \; C \cdot V = 1 \; N \cdot m$

2. The drift velocity of the 2μ particle in this ESP is 7.15×10^{-2} m/s, which is ~15 times of its tangential velocity in a cyclone (4.91×10^{-3} m/s); and it is ~300 times of its terminal velocity due to gravity (2.41×10^{-4} m/s).

Inertial impaction. As a gas stream moves toward a stationary object (e.g., *collection target* in a particulate removal device), it will tend to deflect its flow path around the object (Figure 6.10). For particles in the gas stream, smaller particles having less inertia are likely to move around the object with the gas stream and continue in the path of the gas stream. *Inertia* is a property of matter by which it remains in its existing state of rest or in a constant speed in the line of action. Particles with sufficient inertia will be displaced across the streamlines of the gas stream and move toward to the collection target and get collected/captured.

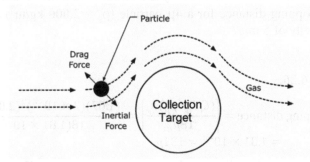

Figure 6.10 - Internal impaction (Modified from EPA, 2012)

For particles having $Re_p < 1$, the effectiveness of impaction can be related to the *inertial impact number* which is defined as:

$$\Psi_I = \left[\frac{(C_c D_p{}^2 \rho_p) v_p}{18\mu_g}\right] \div D_c \quad (6.26)$$

where v_p = particle velocity relative to the collection target and D_c = diameter of the collection target. It should be noted the terms inside the parenthesis in Eq. 6.26 is actually the square of the aerodynamic diameter (see Eq. 6.17). The terms inside the square bracket is often termed *Stokes stopping distance* which represents the distance a particle can travel before it is stopped by viscous friction when the gravitational force is ignored.

The inertial impact number is also called *separation number* or *Stokes number* in literature and sometimes in literature it has a value twice of that shown in Eq. 6.26. As the value of the inertial impact number increases, particles have a greater tendency to move onto the collection target and get collected. On the other hand, the particles will stay on the gas streamlines and move around the target if this parameter approaches zero. One may think that a larger target will yield a more effective impaction. Actually as shown in Equation 6.26, the inertial impact number, which is related to the effectiveness of impaction, is inversely proportional to the diameter of the collection target. Smaller collection targets will be more effective with regards to impaction.

Example 6.9 Stokes stopping distance

Referring to Example 6.4, if the 2µ particle is ejected at a velocity of 5 m/s, how far will it travel before stopped by the viscous friction? What would be the Stokes' stopping distance for a 4µ particle ($\rho_p = 2{,}000$ kg/m^3) ejected at the same velocity of 5 m/s?

Solution:

(a) From Eq. 6.26,

$$\text{Stokes stopping distance} = \left[\frac{(C_c D_p{}^2 \rho_p) v_p}{18\mu_g}\right] = \frac{(1.07)(2 \times 10^{-6})^2 (2{,}000)(5)}{18(1.81 \times 10^{-5})}$$

$$= 1.31 \times 10^{-4} = 131\mu$$

(b) The Stokes stopping distance is proportional to $D_p{}^2$. Therefore, the Stokes stopping distance of the 4µ particle should be four times of that of the 2µ particle; and it would be 524µ.

Discussion: The Stokes stopping distances for both cases are relatively short, and they imply that the air is very viscous for particles to move within it.

Brownian Motion. For very small particles in a gas stream, they will move randomly resulting from their collisions with fast-moving molecules in the gas stream (Figure 6.11); these movements are termed *Brownian motion.* The effect of Brownian motion on particle collection efficiency is related to *diffusional collection parameter* (Ψ_D) which is defined as (EPA, 2012):

$$\Psi_D = D_p \div (D_c v_p) = \left[\frac{kTC_c}{3\pi\mu_g D_p}\right] \div (D_c v_p) \tag{6.27}$$

where D_p = particle diffusivity, k = Boltzmann constant (1.38×10^{-23} kg-m^2/s^2-K) and T = absolute temperature (in K). As shown, particle diffusivity increases with temperature, and decrease with the size and the viscosity of the gas stream.

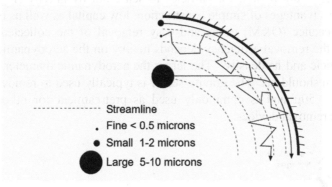

Streamline ---------
. Fine < 0.5 microns
• Small 1-2 microns
⬤ Large 5-10 microns

Figure 6.11 - Particle motion vs. gas streamlines (EPA, 2012)

As Ψ_D increases, the tendency for particles getting collected by Brownian motion increases. Smaller particle size (D_p), collection target (D_c), and relative particle velocity (v_p) values will lead to larger Ψ_D values (i.e., better collection by diffusion). However, Brownian motion is only important for particles having $D_p < 0.3\mu$ and it is only responsible for slight increases in the overall collection efficiency of a particulate removal devices in this size range. In other words, particle diffusion is not a significant particle removal mechanism for most of the control devices (EPA, 2012).

Example 6.10 Particle Diffusivity
Referring to Example 6.4, what is the particle diffusivity of the 2μ particle?

Solution:

From Eq. 6.27,

$$D_p = \left[\frac{kTC_c}{3\pi\mu_g D_p}\right] = \frac{(1.38 \times 10^{-23})(293)(1.07)}{3\pi(1.81 \times 10^{-5})(2 \times 10^{-6})} = 1.19 \times 10^{-11}\, m^2/s$$

Discussion: The diffusivity of this 2μ particle, 1.19×10^{-11} m²/s, is much smaller than diffusivities of gases in air (~10^{-5} m²/s), or those of solutes in water (~10^{-9} m²/s).

6.3 Gravity Settling Chamber

Gravity settling chambers (or *gravity settlers*) are essentially expansion chambers in which the gas flow velocity is reduced, through expansion into a large space, to allow time for particles to settle out by gravity (Figure 6.12). They have advantages of simple construction, low capital as well as operational and maintenance (O&M) costs, and easy removal of the collected particles. However, the removal efficiency depends mainly on the aerodynamic diameter of the particle and its density. The larger the aerodynamic diameter, the better the removal should be. The gravity settler is typically used to remove particles having D_p >50μ and is commonly used as pretreatment for other types of particulate removal devices.

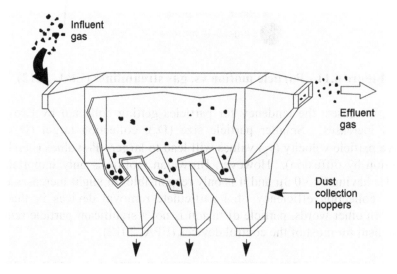

Figure 6.12 - Schematics of a horizontal-flow gravity settler (Modified from EPA, 2012)

138

6.3.1 Principles of gravity settling

A gravity settling chamber is basically a long horizontal rectangular chamber (Length × Width × Height). The cross-sectional area of the chamber (W × H) is much larger than that of the inlet duct so that the gas stream will expand and move slowly through the chamber to allow particles to find time to move downward to the bottom of the chamber by gravity (Figure 6.13).

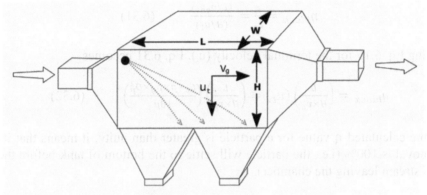

Figure 6.13 - Particle settling in a gravity settler (modified from EPA, 2012)

Assuming the gas velocity (v_{gas}) is uniform through the chamber, it can be readily determined as:

$$v_{gas} = \left. Q_{gas} \middle/ (W \times H) \right. \qquad (6.28)$$

where Q_{gas} = flow rate of the gas stream. For a particle entering the chamber, it will have a horizontal velocity (u_h) which is the same as v_{gas} and a vertical settling velocity which is its terminal velocity (u_t). The residence time of the gas stream in the chamber (τ) can be readily calculated as:

$$\tau = \frac{L \times W \times H}{Q} = \frac{L}{Q/(W \times H)} = \frac{L}{v_{gas}} \qquad (6.29)$$

The time needed for a particle to settle to the bottom of the tank (t_s) can be found by dividing the height of chamber with the terminal velocity:

$$t_s = \frac{H}{u_t} \qquad (6.30)$$

A particle will be considered removed if it can settle to the bottom before the gas exits the chamber. For a block flow pattern, we assume that the particles are distributed uniformly across the cross-section at the inlet (i.e., some particles will enter at the very top of the chamber while the other will enter closer to the bottom of the chamber). We also assume that there is no mixing or interaction among the particles in both the transverse and the longitudinal directions of the gas flow. For this gas flow pattern, removal efficiency (η_{block}) for particles of a specific terminal velocity is essentially the ratio of the gas residence time (τ) and the traveling time needed for these particles to settle to the bottom of the tank (t_s); that is:

$$\eta_{block} = \frac{\tau}{t_s} = \frac{(L/v_{gas})}{(H/u_t)} \qquad (6.31)$$

Using Eq. 6.15 for the terminal velocity (u_t), Eq. 6.31 becomes

$$\eta_{block} = \left(\frac{L}{H \times v_g}\right)(u_t) = \left(\frac{L}{H \times v_g}\right)\left(\frac{C_c \times g \times \rho_g \times D_p^2}{18\mu}\right) \qquad (6.32)$$

If the calculated η value for a particle is greater than unity, it means that its removal is 100% (i.e., the particle will settle to the bottom of tank before the gas stream leaving the chamber).

For an extreme case that is totally opposite to the block flow model, we assume that the gas stream is totally mixed in the transverse direction, but no mixing in the longitudinal direction. For this case, the mixing in the transverse direction would reduce the removal efficiency, and the removal efficiency based on the mixed flow model (η_{mixed}) can be found from that of the block flow model (η_{block}) as:

$$\eta_{mixed} = 1 - \exp[-\eta_{block}] = 1 - \exp\left[-\left(\frac{L}{H \times v_g}\right)\left(\frac{C_c \times g \times \rho_g \times D_p^2}{18\mu}\right)\right] \qquad (6.33)$$

For real-life situations, the gas flows will be turbulent to some extent, and the mixing in the transverse direction will reduce the removal efficiency. On the other hand, the mixing in the longitudinal direction will allow the particles to move both upstream and downstream with insignificant impacts on the overall removal efficiency. Consequently, the actual removal efficiency should be between the values predicted the block and the mixed flow models (i.e., Equations 6.32 and 6.33).

6.3.2 Gravity settling systems and their components

As shown in Eq. 6.31, the removal efficiency is inversely proportional to the height of the chamber (H). This makes sense because the particles would not need to settle a long distance to reach the bottom of the chamber for a smaller H. Therefore, there are multi-tray settling chambers available, in which there are several horizontal trays within the rectangular chamber so that the settling distances are much shorter (Figure 6.14).

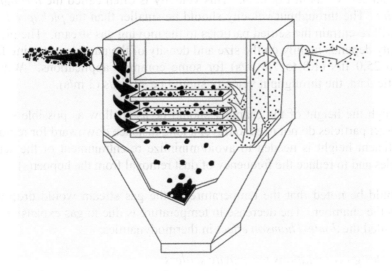

Figure 6.14 - Schematics of a multi-tray gravity settler (EPA, 2012)

To minimize the mixing in the transverse direction of the gas stream flow, inlet and outlet baffles are often added to guide the flow so it will be more parallel in the axial/longitudinal direction.

As shown in both Figures 6.12 and 6.13, in addition to the rectangular portion for gas expansion, the collection hopper located at the bottom of the settling chamber is another key component. The hoppers are often designed with sealing valves and the collected particles need to be emptied frequently to avoid dust re-entrainment. The potential *dust re-entrainment* is a constraint to the spacing between the trays for multi-tray gravity settlers (EPA, 2012).

6.3.3 Factors affecting particle removal by gravity settling

The terminal velocity derived from the balance of the gravitational, buoyancy, and drag forces is the key parameter for design of a settling chamber. The gas residence time (τ), which is the ratio of the chamber volume and the gas flow rate, should be long enough to allow sufficient time for target particles to settle. With that, the design variables include the length, width, height of the rectangular portion of the chamber.

For a given flow rate, the height and the width of the chamber (i.e., the cross-sectional area) will determine the velocity at which the gas moves through the chamber, v_g, as in Eq. 6.28. This velocity is often called the *throughput velocity*. The throughput velocity should be smaller than the *pickup velocity* that will re-entrain the settled particles to the moving gas stream. The pickup velocity depends mainly on the size and density of the particles, ranging from 5.8 to 25.0 ft/s (1.8 to 7.6 m/s) for some common applications. Without specific data, the throughput velocity should be <10 ft/s (3 m/s).

Although the height of the chamber should be as shallow as possible so that the target particles do not need to travel a long distance downward for removal, a sufficient height is needed to avoid/minimize re-entrainment of the settled particles and to reduce the frequency of dust removal from the hopper(s).

It should be noted that the temperature of the gas stream would drop as it enters the chamber. The decrease in temperature is due to gas expansion and it is called the *Joule-Thomson effect* in thermodynamics.

6.3.4 Design calculations for gravity settlers

For selection of any pollution control device, its removal efficiency is often one of the major considerations. Sometimes, we may need to put two or more devices (they may be same or different) in series to increase the overall removal efficiency. In other times, we may need to operate control devices in parallel to accommodate needs of maintenance or fluctuations in the influent flow rates. *Removal efficiency* (η) is the fraction/percentage of the influent pollutants that is removed by the device, while penetration (p) is the fraction/percentage of the influent pollutants that is not removed (i.e., escape from the control device). The relationship between them is very straightforward:

$$Penetration\ (p) = [1 - Removal\ efficiency\ (\eta)] \qquad (6.34)$$

If we employ removal devices in series, using penetration will make the calculation of the overall removal efficiency easy. For example, if we have two control devices in series with removal efficiencies of η_1 and η_2, respectively, the removal efficiency can be found by:

$$\eta_{overall} = 1 - (1 - \eta_1)(1 - \eta_2) = 1 - (p_1)(p_2) \qquad (6.35)$$

The example below illustrates the design calculations for a horizontal gravity settling chamber.

Example 6.11 Gravity settling chamber

A gravity settling chamber (L × W × H = 10 m × 4 m × 3m) is used to treat a gas stream (Q = 20 m^3/s; T = 20 °C; P = 1 atm) which contains particles of three distinct sizes (D$_p$ = 2, 50 and 100 μ). The densities of all particles is the same, 2,000 kg/m^3. The mass fractions of the 2, 50, 100μ particles are 5, 25, and 70%, respectively. Determine

(a) the throughput velocity of the gas stream;

(b) the terminal velocities of the particles;

(c) the removal efficiency for each particle size; and

(d) the overall removal efficiency of all particles.

Solution:

(a) From Eq. 6.28,

$$Throughput\ velcoity = v_{gas} = {^{20}}/{(10 \times 4)} = 0.5\ m/s$$

(b) $(u_t)_{2\mu} = 2.41 \times 10^{-4}$ m/s (from Example 6.4)

For the 50μ and 100μ particles, their Cunningham slip correction factors are essentially equal to one. Since the terminal velocity is proportional to D$_p^2$, we can find their terminal velocities by using that of the 2μ particles.

$(u_t)_{50\mu} = [(u_t)_{2\mu}] \times (1/1.09)(50/2)^2 = [(2.41 \times 10^{-4})/1.09](25)^2 = 0.138$ m/s

$(u_t)_{100\mu} = [(u_t)_{50\mu}] \times (100/50)^2 = (0.138)(4) = 0.552$ m/s

(c) From the 2μ particles and Eq. 6.31:

$$\eta_{block} = \left(\frac{L}{H \times v_g}\right)(u_t) = \left(\frac{10}{3 \times 0.5}\right)(2.41 \times 10^{-4}) = 0.0016$$

From Eq. 6.32,

$$\eta_{mixed} = 1 - \exp[-\eta_{block}] = 1 - \exp(-0.0016) = 0.0016$$

The removal efficiencies for 50μ and 100μ particles can be determined using the approach above by inserting the corresponding u_t value. The results are tabulated below for comparison.

Particle size (μ)	Terminal velocity (m/s)	η (Block flow)	η (Mixed flow)
2	0.000241	0.0016	0.0016
50	0.138	0.92	0.60
100	0.552	1.00	0.97

As expected, the removals using the mixed flow model are smaller than those of the block flow model. The actual removals should be between the values predicted by these two models. This device can effectively remove particles having $D_p > 50μ$.

(d) Using $η_{mixed}$, as the worst-case scenario to estimate the total mass removal:

Basis: total input particle mass = 100 kg

Particle size (μ)	Inlet mass (kg)	η	Penetration	Outlet mass (kg)
2	5	0.0016	0.9984	4.992
50	25	0.60	0.4	10
100	70	0.97	0.03	2.1
Total	**100**			**17.092**

Total removal = $1 - p = 1 - (17.092/100) = 82.9\%$

Discussion:

1. Since the question is to find the extent of removal, not the exact mass in the outlet, "basis" for calculations can be arbitrarily chosen to facilitate the calculations. Although a total mass of 100 kg is unreasonably high, it will not affect the results. However, 100 g would appear to be more reasonable.

2. As shown, this gravity settling chamber can remove lots of large particles and the majority of the inlet mass, but the removal of 2μ particles is insignificant. If they need to be removed to a larger extent, other types of particle removal devices such as fabric filters, ESPs, or wet scrubbers should be employed.

6.4 Cyclones

Cyclones have a cylindrical shape which causes the gas stream moving in high-velocity turns that only gas molecules and tiny particles with small inertia can move with it. Large inertia of large particles will prevent them from making the turns so to hit the inner wall of the cyclone and get collected.

Figure 6.15 illustrates the *fractional efficiency curves* of cyclones. Cut size (defined later) of a simple cyclone is about 5μ. If removal of finer particles is needed, cyclones are often used as pre-collectors to other types of device capable of removing finer particles such as fabric filters and electronic precipitators.

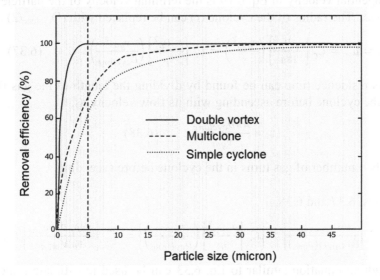

**Figure 6.15 - Fractional efficiency curves of cyclones
(modified from EPA, 2012)**

6.4.1 Principles of cyclones

As discussed in the previous section for the gravity settling chambers, the removal efficiency for a particle depends on the ratio of the gas residence time (τ) and the time needed for the particle to reach the bottom of the chamber (t_s). For cyclones, the concept is essentially the same, except the particle does not need to travel to the bottom of the cyclone to get removed. Instead, it only needs to hit the wall of the cyclone to get removed when the gas is circulating downward. The gas stream then spins upward to the gas outlet carrying the uncollected particles with it. Assuming the block flow model, the maximum distance that a particle needs to travel to the wall is the width of the rectangular inlet (W_{inlet}), versus the height (H) of the settling chamber for gravitational settling. In the meantime, the length of the gas travel in which the particles having opportunity to hit the wall is $N(\pi D_{cyclone})$, compared to the length of the gravity settling chamber (L). N is the number of turns that the gas stream makes before it spins upward and $D_{cyclone}$ is the diameter of the cyclone. With these, the removal efficiency can be derived, similar to Eq. 6.32, as:

$$\eta_{block} = \frac{\tau}{t_{wall}} = \frac{(N\pi D_{cyclone}/v_{gas})}{(W_{inlet}/u_t)} = \left[\frac{N\pi D_{cyclone}}{(W_{inlet})(v_{gas})}\right] \times u_t \qquad (6.36)$$

where t_{wall} = time needed for the particle to travel to the wall.

The tangential velocity in Eq. 6.19 is the terminal velocity of the particles in a cyclone. The radius of the cyclone (r) can be replaced with ($D_{cyclone}/2$) as,

$$u_t = u_p = C_c\left[\frac{\rho_p D_p{}^2}{18\mu}\right]\left(\frac{v_{gas}{}^2}{r}\right) = C_c\left[\frac{\rho_p D_p{}^2}{9\mu}\right]\left(\frac{v_{gas}{}^2}{D_{cyclone}}\right) \qquad (6.37)$$

The gas residence time can be found by dividing the length of the gas flow inside the cyclone before ascending with its flow velocity as:

$$\tau = \frac{N\pi D_{cyclone}}{v_{gas}} \qquad (6.38)$$

where N = number of gas turns in the cyclone before ascending.

From Eqs. 6.37 and 6.38,

$$\eta_{block} = \left[\frac{N\pi D_{cyclone}}{(W_{inlet})(v_{gas})}\right] \times \left\{C_c\left[\frac{\rho_p D_p{}^2}{9\mu}\right]\left(\frac{v_{gas}{}^2}{D_{cyclone}}\right)\right\} = C_c\left[\frac{\pi N \rho_p D_p{}^2 v_{gas}}{9\mu W_{inlet}}\right] \qquad (6.39)$$

Although an equation similar to Eq. 6.33 can be used to estimate removal efficiencies based on the mixed flow model, the block flow model has been more widely used because its predictions are reasonably good (de Nevers, 2000).

As shown in Eq. 6.39, the removal of a specific particle by a cyclone is proportional to the density of the particle (ρ_p) and its diameter to the power of two ($D_p{}^2$) and inversely proportional to the viscosity of the gas stream (μ). The removal is also proportional to the number of turns (N) and inversely proportional to the width of the rectangular inlet (W_{inlet}); these two variables (i.e., N and W_{inlet}) depend on the physical configuration of the cyclone. The removal is also proportional to the velocity of the gas stream (v_{gas}). For a given gas flow rate (Q_{gas}), the v_{gas} will be larger for a rectangular inlet having a smaller cross-section ($H_{inlet} \times W_{inlet}$). If the aspect ratio of the rectangular inlet is fixed, the v_{gas} would be inversely proportional to $(W_{inlet})^2$, and the removal will then be inversely proportional to $(W_{inlet})^3$ accordingly as:

$$\eta_{block} \sim \left[\frac{v_{gas}}{W_{inlet}}\right] \sim \left[\left(\frac{Q_{gas}}{L \times W_{inlet}}\right)/W_{inlet}\right] \sim \left[\frac{Q_{gas}}{L \times W_{inlet}^2}\right] \sim \left[\frac{Q_{gas}}{W_{inlet}^3}\right] \quad (6.40)$$

Although a larger v_{gas} or a smaller W_{inlet} will yield a higher removal efficiency, it will incur a significantly larger head-loss and energy costs.

6.4.2 Cyclones and their components

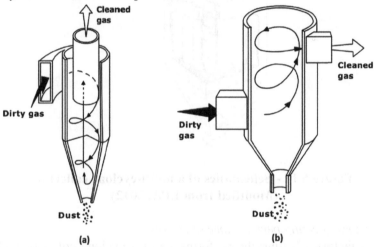

Figure 6.16 - Large-diameter Cyclones – (a) top inlet; (b) bottom inlet (Modified from EPA, 2012)

The cyclones for industrial applications can be classified into two groups: (i) large-diameter cyclones and (ii) small-diameter multi-clones. Figure 6.16 shows schematic of large-scale cyclones. As shown, they have a cylindrical portion on the top and a conical portion at the bottom. Dirty gas can come from the top or the bottom portion of the cylindrical portion. The gas enters the rectangular inlet, normally the height of the inlet is twice of its width. The inlet is typically arranged tangentially to the cylindrical portion of the cyclone. After a few turns, the gas stream gets cleaner and exits from the top while the collected particles are removed from the conical bottom of the cyclone. Diameters of the large-diameter cyclones range from approximately one foot (0.3 m) to >12 feet (3.6 m) and typical operating head-loss is from 2 to 4 in-H_2O (0.5 to 1.0 kPa).

A mutli-cyclone collector is a group of small-diameter cyclones, typically 6 to 12 inches (0.15 to 0.3 m) in diameter (Figure 6.17). It has better particulate removal capability than large-diameter cyclones (see Figure 6.15), but is more expensive and would incur a larger head-loss, > 4-in H$_2$O (1.0 kPa) (EPA, 2003).

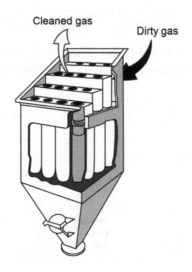

Figure 6.17 - Schematics of a multi-cyclone collector (Modified from EPA, 2012)

6.4.3 Factors affecting performance of cyclones
Important factors affecting the performance of a cyclone collector include the size and density of the particles, the gas velocity through the cyclone, the residence time of the gas, and the configurations of the cyclone (i.e., dimensions of the body as well as those of the inlet).

6.4.4 Design calculations for cyclones
Cut diameter (or cut size), D$_{cut}$, has been widely used in describing the size ranges of particles that can be collected by a particulate removal device. It is defined as the diameter of the particles collected with 50% efficiency. The particles with D$_p$ > D$_{cut}$ will be collected by more than 50% and vice versa.

By letting η_{block} = 50% and D$_p$ = D$_{cut}$, Eq. 6.39 can be rearranged to:

$$D_{cut} = \left(\frac{9W_{inlet}\mu}{2\pi C_c N v_{gas}\rho_p}\right)^{0.5} \cong \left(\frac{9W_{inlet}\mu}{2\pi N v_{gas}\rho_p}\right)^{0.5} \qquad (6.41)$$

Since the typical D_{cut} values for cyclones are a few microns and the Cunningham slip correction factors (C_c) for particles in this size range is essentially equal to 1, the C_c is often omitted in Eq. 6.41.

Example 6.12   Residence time and cut diameter of a cyclone

A cyclone has a diameter of 1 m and a rectangular inlet (H = 0.4 m and W = 0.2 m). The gas flow rate is 2.4 m³/s. Determine

(a) the gas flow velocity;

(b) the gas residence time in the cyclone for particulate removal (N = 5); and

(c) the cut diameter, if the density of the particles is 2,000 kg/m³.

Solution:

(a) $v_{gas} = (Q_{gas})/(H \times W_{inlet}) = (2.4)/[(0.4)(0.2)] = \underline{30 \text{ m/s}}$

(b) From Eq. 6.37,

$\tau = (N \pi D_{cyclone})/v_{gas} = [5\pi(1)]/(30) = \underline{0.52 \text{ s}}$

(c) From Eq. 6.41,

$$D_{cut} = \left(\frac{9W_{inlet}\mu}{2\pi N v_{gas}\rho_p}\right)^{0.5} = \sqrt{\frac{9(0.2)(1.81 \times 10^{-5})}{2\pi(5)(30)(2,000)}} = 4.2 \times 10^{-6}m = 4.2\mu$$

Discussion:

1. The gas residence for this operation is only 0.52s, which is much shorter than that of a gravity settling chamber.

2. The cut size is 4.2μ, which falls into the range of typical cyclones.

Example 6.13   Cut diameters of two cyclones

Referring to Example 6.12, if we feed the same air stream (same Q, particle loading, PSD) to a cyclone of the same design, except all the dimensions are half as big, what will be the new cut diameter (using the block flow model)? What are the implications?

149

Solution:

(a) Assuming N stays the same, from Eq. 6.41:

$$\frac{D_{cut,new}}{D_{cut,old}} = \sqrt{\left(\frac{W_{inlet,new}}{W_{inlet,old}}\right)\left(\frac{v_{gas,old}}{v_{gas,new}}\right)} = \sqrt{\left(\frac{1}{2}\right)\left(\frac{1}{4}\right)} = 0.35$$

$D_{cut,new} = (4.2)(0.35) = \underline{1.5\mu}$

(b) D_{cut} will become smaller (from 4.2 to 1.5μ). However, because the dimensions are reduced into half, the gas velocity will be four times larger. The corresponding head-loss would much higher (discussed later in this section)

Since a typical gas stream should contain particles of different sizes. An empirical equation (Eq. 6.42 below) has been developed to determine the removal efficiency (η) for each size, using the ratio of D_p and D_{cut} (i.e., D_p/D_{cut}). The equation has been found to fit the actual performances of cyclones well.

$$\eta = \frac{\left(D_p/D_{cut}\right)^2}{1+\left(D_p/D_{cut}\right)^2} \qquad (6.42)$$

Using Eq. 6.42, a collection efficiency can be generated as shown in Figure 6.18. The removal efficiency is equal to 50%, when the D_p/D_{cut} ratio = 1.

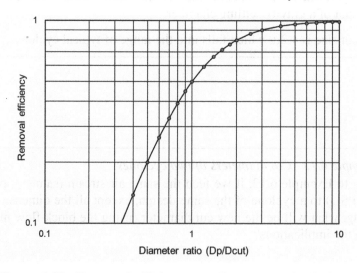

Figure 6.18 - Removal efficiency vs. diameter ratio (D_p/D_{cut})

Example 6.14 Removal efficiencies of a cyclone

A particle stream has 50% by wt. of particles having $D_p = 1\mu$ and the other 50% (by wt.) having $D_p = 25\mu$. If we pass this stream through a cyclone with $D_{cut} = 5\mu$, how much of the particles (by wt.) will be removed by this cyclone?

Solution:

(a) Using Eq. 6.42 to calculate the removal efficiencies:

$$\eta_{1\mu} = \frac{\left(1/5\right)^2}{1 + \left(1/5\right)^2} = 0.038; \quad \eta_{25\mu} = \frac{\left(25/5\right)^2}{1 + \left(25/5\right)^2} = 0.96$$

(b) $(\eta_{removal})_{total} = (0.038)(50\%) + 0.96(50\%) = 49.9\%$

Discussion:

1. With $D_{cut} = 5\mu$, the removal for 25μ particles approaches 100% while that for 1μ particles is <4%.

2. The total removal comes mainly from removal of the large particles.

Pressure drop is an important operational parameter for cyclones. As shown in Eq. 6.41, the cut size (D_{cut}) is inversely proportional to the square root of the gas velocity (v_{gas}). It means that a larger v_{gas} will reduce D_{cut} and improve the removal efficiency of a cyclone. However, a larger V_{gas} will incur a larger pressure drop (or head-loss), ΔP, that is the difference between the inlet pressure (P_{in}) and the outlet pressure (P_{out}). A larger ΔP demands a more powerful air movement device and incurs more energy costs to move the gas stream through the cyclone. The pressure drop can be estimated based on the quantity of ($\rho_{gas}v_{gas}^2/2$), which is termed as *velocity head* as:

$$\Delta P = (P_{in} - P_{out}) = K \left(\frac{\rho_{gas}v_{gas}^2}{2}\right) \quad (6.43)$$

If you have taken fluid mechanics, this approach is commonly used to estimate the head-loss through a fluid transportation system. There are different K values for different components such as elbows, valves, flow meters, etc. Since each cyclone has its own specific design so it will have its own specific K, which can be derived from experimental data. Since K is an empirical value, the reported K values are often tied to the units of ΔP, ρ_{gas} and v_{gas} used in equation. According to EPA (2012), typical K values range from 0.026 to 0.048, if ΔP is in inches of water, ρ_{gas} in lb/ft^3 and v_{gas} in ft/s. de Nevers (2000) stated that most cyclones have pressure drops of about 8 velocity heads (i.e., K = 8).

Example 6.15 Pressure drop through a cyclone

Estimate the head-loss through a cyclone if the gas velocity is 50 ft/s (15.24 m/s) and T = 68 °F (20 °C).

Solution:

From Example 2.2, ρ_{air} = 0.075 lb/ft^3 (1.21 g/L) @ T = 68 °F (20 °C)

(a) Using the approach of EPA (2012) and assuming K = 0.048 (the high end of the range):

$$\Delta P = (0.048)\left(\frac{(0.075)(50)^2}{2}\right) = 4.5 \ in - H_2O$$

(b) Using the approach of de Nevers (2000):

$$\Delta P = (8)\left(\frac{1}{2}\right)\left(0.075 \ \frac{lb_m}{ft^3}\right)\left(50 \ \frac{ft}{s}\right)^2\left(\frac{lb_f}{32.2 lb_m \cdot \frac{ft}{s^2}}\right) = 23.3 \ \frac{lb_f}{ft^2}$$

$$= \left(23.3 \ \frac{lb}{ft^2}\right)\left(\frac{ft^2}{144 in^2}\right)\left(\frac{407 \ in - H_2O}{14.7 \ psi}\right) = 4.48 \ in - H_2O$$

Discussion: A K value of 8 as suggested by de Nevers (2000) is essentially equivalent to 0.048 of EPA (2012) with specific units for three parameters.

Example 6.16 Pressure drop of a cyclone (SI units)

Referring to the cyclone in Example 6.15, estimate the pressure drop using SI units.

Solution:

Assuming K = 8, insert the given values into Eq. 6.43:

$$\Delta P = (8)\left(\frac{1}{2}\right)\left(1.21 \frac{kg}{m^3}\right)\left(15.24 \frac{m}{s}\right)^2\left(\frac{N}{kg \cdot \frac{m}{s^2}}\right) = 1,126 \ N/_{m^2}$$

Example 6.17   Electricity cost of a cyclone (SI and the US customary units)

Referring to the cyclone in Examples 6.15 and 6.16, if it operates continuously and the blower efficiency is 85%, estimate the annual electricity cost. The flow rate is 10 m³/s (353 ft³/s) and electricity cost is $0.20/kW-hr.

Solution:

(a) Using the SI units:

$$Power = \frac{(10\frac{m^3}{s})(1{,}126\frac{N}{m^2})}{0.85} = 13{,}247\ \frac{N \cdot m}{s} = 13{,}247\ W$$

$$\$ = (13{,}247W)(365\ d)\left(24\frac{hr}{d}\right)\left(\frac{kW}{1{,}000\ W}\right)(\$0.20\ kW - hr) = \$23{,}210$$

(b) Using the US Customary units:

$$Power = \frac{(353\frac{ft^3}{s})(23.3\frac{lb_f}{ft^2})}{0.85} = \left(9{,}676\ \frac{ft \cdot lb_f}{s}\right)\left(\frac{745.7\ W}{550\ ft \cdot lb_f/s}\right)$$
$$= 13{,}120\ W$$

6.5 Electrostatic Precipitators

An electrostatic precipitator (ESP) can be envisioned as a large number of parallel plates with discharge electrodes located in the center plane between the plates. The plates are electrically-grounded so that electrical fields are established between the discharge electrodes and the plates. The gas stream moves between the plates and the particles in the gas stream will acquire negative charges from the electrons discharged from the electrodes or from the negative gas ions formed from collisions with the negative electrons. The particles carrying the negative charges will then move toward the plates. Once they hit the plates, they are considered collected. Removal of particles by ESPs is based on applying a surface force (i.e., electrostatic), instead of a body force (e.g., gravitational, centrifugal) to fine particles. The process is applicable for solid and liquid particles.

6.5.1 Principles of electrostatic precipitators

Similar to terminal velocity for particles in gravity settling chambers and tangential velocity of particles in cyclones, drift velocity (w) is how fast a particle will move toward a collection plate of an ESP. To determine the removal efficiency, let's go back to the simplified system - two identical plates (with length L and a height H) are parallel to each other with a separation distance of S. Several electrodes are placed in the center plane between these two plates. The gas stream will flow through the channel formed by these parallel plates.

Now let's work the right (or the left) half of the channel. The gas flow rate (Q) through this half of the channel will be the multiplication product of the cross-sectional area and the gas velocity (v_{gas}) as:

$$Q = \left(\frac{S}{2} \times H\right) \times v_{gas} \qquad (6.44)$$

The collecting area (A) will be the area of one single plate, which is:

$$A = L \times H \qquad (6.45)$$

Similar to the approach we made for the gravitational settling chambers with the block flow pattern, removal efficiency (η_{block}) for particles of a specific drift velocity (w) is essentially the ratio of the gas residence time (τ) and the traveling time needed for these particles to reach the collection plate (t), that is:

$$\eta_{block} = \frac{\tau}{t} = \frac{(L/v_{gas})}{\left(\frac{S}{2}\right)/w} = \frac{w(L \times H)}{\left(\frac{S}{2} \times H\right) \times v_{gas}} = \frac{wA}{Q} \qquad (6.46)$$

Similar to gravitational settling chambers, the removal efficiency of the mixed flow mode (η_{mixed}) can be derived as:

$$\eta_{mixed} = 1 - exp[-\eta_{block}] = 1 - \exp\left[-\frac{wA}{Q}\right] \qquad (6.47)$$

Eq. 6.47, instead of Eq. 6.46, is widely used in practices for conceptual design and analysis of ESPs. It has a name, *Deutsch-Anderson equation*. The ratio of the collection surface and the gas flow rate (i.e., A/Q) is often called *specific collection area* (SCA). It should be noted that any consistent set of units can be used for w and (A/Q) as long as "wA/Q" is dimensionless.

As mentioned earlier, there are two mechanisms with regards to particle charging: contact charging and diffusional charging. The combined effect of these two mechanisms creates an efficiency-particle size curve as shown in Figure 6.19. The efficiencies are high for particles having D_p >~1μ because of the effectiveness of contact charging increases with particle size. On the other hand, removal efficiencies for particles having D_p <~1μ increase as the particle size decreases because of the increase in diffusional charging for smaller particles. The removal efficiencies for particles in the size range between 0.1 to 1μ are lower because of the size-dependent limitations of both charging mechanisms (EPA, 2012).

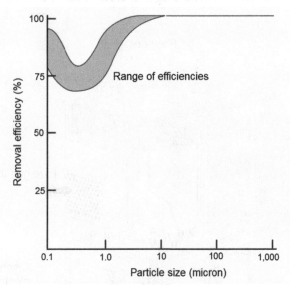

**Figure 6.19 - Factional efficiency curve of ESPs
(Modified from EPA, 2012)**

6.5.2 ESP systems and their components
ESPs can be used to collect solid as well as liquid particles. To meet different needs of industrial applications, there are three categories of ESPs: (i) dry, negative corona, (ii) wet, negative corona, and (iii) wet, positive corona.

Dry, negative corona type ESPs are most commonly used. The collected solids are removed from the collection plates as a "dry" material. They are termed "negative corona" because the particles are forced to move from a negatively-charged zone to the electrically-grounded collection plates. As shown in Figure 6.20, the influent gas is guided through perforated screen to

155

go through passages formed by many parallel collection plates at a velocity between 3 to 6 ft/s (1 to 2 m/s). The typical spacing between two plates is 12 to 18 inches (0.3 to 0.45 m). High voltage discharge electrodes are centered between each pair of the plates. The electrodes can have small diameters as shown in Figure 6.21. The high voltages applied to the electrodes create a negative corona that would charge most of the particles negatively. The discharge electrodes are divided into fields that are in series and the ESPs can also be divided into separate chambers that are in parallel by a solid wall. It appears that the system shown in Figure 6.20 has two chambers and each chamber has three fields. The dust layer accumulated on the collection plates are periodically removed by mechanical hammers, called *rappers*. They fall into hoppers by gravity as particle agglomerates and/or as dust layer sheets (EPA, 2012).

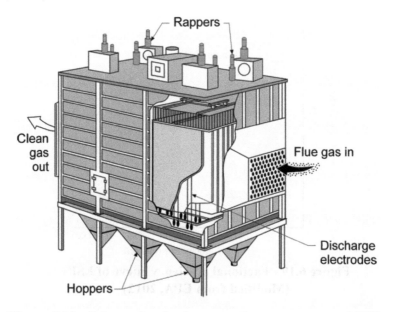

Figure 6.20 - Schematic of a dry, negative corona type of ESP (EPA, 2012)

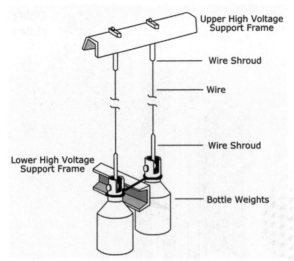

Figure 6.21 - Wire-type discharge electrodes
(EPA, 2012)

Wet, negative corona type ESPs are used where mists or fogs need to be removed or when solid particles to be removed have undesirable electrical or physical properties so that a washing system, instead of rappers, is used to remove particles accumulated on the collection plates (EPA, 2012).

Wet, positive corona type ESPs are used to remove organic droplets or mists from small industrial applications. The positive, instead of negative, voltages are applied to the discharged electrodes which are housed in a pre-ionizer section, ahead of the electrically-grounded collection plates downstream. Since the ESPs only collect liquid particles, no rappers or washing liquid distribution systems are needed (Figure 6.22). The plates are cleaned periodically (EPA, 2012).

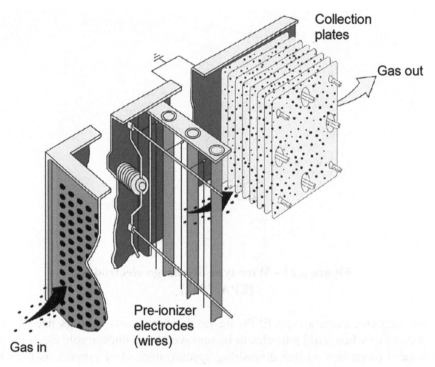

Collection plates

Gas out

Pre-ionizer electrodes (wires)

Gas in

Figure 6.22 - Wet, positive corona type ESP
(Modified from EPA, 2012)

6.5.3 Factors affecting performance of ESPs

From the Deutsch-Anderson equation (Eq. 6.47), the drift velocity (w) and the SCA (the A/Q ratio) are the only apparent variables affecting the performance of an ESP. The larger the SCA, the higher the removal efficiency would be. Therefore, for an ESP with a fixed total surface area of collection plates, lowering the gas flow rate into the treatment zone by reducing its temperature should raise the removal efficiency.

The larger the drift velocity, the corresponding removal efficiency will also be higher because the target particles will move to the collection plates at a faster speed. As shown in Eq. 6.25, the drift velocity is proportional to the particle diameter (D_p) and to the square of the electricity field strength (E). A stronger field will make the drift velocity larger. The electricity field strength is nothing but the gradient of the voltage with distance. It seems to be the simple ratio of the voltage difference, between the discharge electrodes and the electrically-grounded collection plates, and the distance between them. However, in actual operations the cakes accumulate on the collection plates will have impacts on the strength of the electricity field where the particles are being charged.

The voltage on the surface of an electrically-grounded plate is zero; and the electrostatic voltage difference across the dust cake holds the cake on the vertical surface of the collection plate. The electrostatic voltage of the outermost surface of the dust layer, where new particles and gas ions are arriving, can be >10,000V. The ability of electrical charges to move through a dust layer is measured by the resistivity of the dust layer. A commonly-used resistivity unit is Ohm-cm. When the resistivity is lower, the electrons are conducted more readily; the charge difference across the dust layer will be smaller. The resistivity depends on the composition of the cake materials and the operating conditions (e.g., temperature and moisture content).

ESPs typically work best when the resistivity of the dust cake is in the moderate range (5×10^6 to 5×10^{10} Ohm-cm). For low-resistivity dust cakes, they are weakly held. During rapping, many of the particles or their small agglomerates may be released back into the gas stream instead of moving downward to the collection hoppers. It will result in a temporary emission spike, called a *puff*.

If the resistivity of the dust layer is too high, the voltage at the outermost surface would become much larger. Consequently, the voltage difference between the discharge electrode and the cake's outermost surface becomes smaller, so as the electrostatic field strength. The migration speed of the particles toward the collection plate will become slower. The second adverse effect of high-resistivity dust layers is *back corona* which will have adverse impacts on removal efficiency. It is because corona discharges in the gas trapped inside the cake will occur if the voltage drop across the dust layer becomes sufficiently large. Some gas molecules will become positively charged after collision with negative ions. The positively-charged gas ions will move backward toward the discharge electrode. On their way, they will neutralize some of the negative charges on the particles in the dust layer as well as some of the negative ions on the particles approaching the cake. The third adverse impact is that frequency of electrical sparking within the high-resistivity cake layer may become too frequent to be desirable. In addition to all these, it is also difficult to dislodge the dust cakes because they are more strongly held by the stronger electrostatic fields.

Not much one can do about the resistivity of the sludge cake when it is below the moderate/preferred range. For the high-resistivity dust, its resistivity can decrease as the gas temperature drops because of the increased adsorption of vapors which are electrically conductive. Some common chemicals, such as sulfur trioxide and ammonia can also be injected into the gas stream inlet to improve the *hygroscopic* property (i.e., tendency to absorb moisture from the air) of the particles.

6.5.4 Design calculations for ESPs

For conceptual design of an ESP, the drift velocity (w) and SCA need to be specified to estimate the removal efficiency. The drift velocity depends on characteristics of particles (e.g., dielectric constant and particle size) as well as those of the system (e.g., strength of the electrostatic field). According to EPA (2012), typical drift velocity values range from 0.13 to 0.67 ft/s (4.0 to 20.4 cm/s). The SCA value tells the total surface area needed for a given gas flow rate. The present-day SCA vales range from 300 to 1,400 ft² per 1,000 scfm (0.3 to 1.4 min/ft), or 1.0 to 4.6 min/m (EPA, 2012). Although the ranges of w and A/Q are provided here, the proper values to achieve a specific particle removal would depend on many site-, unit-, and particle-specific factors such as PSD and resistivity, aspect ratio of the plates, and electric field strength, to name a few.

Example 6.18 Performance of ESPs

Referring to the 2-μ particles in Example 6.6, if an ESP is to be employed to achieve a 99% removal efficiency,

(a) what would the required surface area of the collection plates?

(b) if the flow rate is doubled, what would be the removal efficiency?

(c) what would be the removal efficiency for 1μ particles [Note: T = 20 °C; Q = 20 m³/s (706 ft³/s); w = 7.15 cm/s]

Solution:

(a) Using Eq. 6.47:

$$\eta = 99\% = 1 - exp\left[-\frac{wA}{Q}\right] = 1 - exp\left[-(7.15\frac{cm}{s})\left(\frac{A}{Q}\right)\right]$$

(A/Q) = 0.644 s/cm = 64.4 s/m = 1.07 min/m

A = (64.4 s/m)(20 m³/s) = 1,288 m²

(b) If the flow rate is doubled, then (A/Q) = (0.644/2) = 0.322 s/cm.

Assuming that w is independent of Q (not a bad assumption), then

$$\eta = 1 - exp[-(7.15)(0.322)] = 90\%$$

(c) Since w is proportional to D_p, the drift velocity of the 1μ particles will be half of that of the 2μ particles. With the same (A/Q), the removal efficiency will be the same as that of (b), 90%, since they have the same [w(A/Q)] value.

Discussion:

1. In this example, drift velocity (7.15 cm/s) and SCA (1.07 min/m) are within the typical ranges. This example illustrates that ESPs are effective in removing fine particles ($D_p = 2\mu$ in this example).

2. Units of w and (A/Q) should match.

3. The efficiency decreases from 99% to 90% if the flow rate doubles.

4. The efficiency of 1μ particles are lower, 90%, but still relatively high, when compared to that of typical cyclones.

To take a more conservative approach, some would like to use the *modified Deutsch-Anderson equation* as:

$$\eta == 1 - \exp\left[-\left(\tfrac{wA}{Q}\right)^k\right] \qquad (6.48)$$

where k is an empirical constant, ranging from 0.4 to 0.6.

Example 6.19   The modified Deutsch-Anderson equation

Referring to Example 6.18, use the modified Deutsch-Anderson equation (k = 0.5) to estimate the removal efficiencies for the 2μ and the 1μ particles by an ESP with A/Q = 0.644 s/cm.

Solution:

(a) The drift velocity of the 2μ particles = 7.15 cm/s.

Using Eq. 6.48:

$$\eta = 1 - exp[-(7.15 \times 0.644)^{0.5}] = 88.3\%$$

(b) The drift velocity of the 1μ particles = (7.15/2) = 3.575 cm/s.

$$\eta = 1 - exp[-(3.575 \times 0.644)^{0.5}] = 78.1\%$$

Discussion: As expected, the estimated removal efficiencies are lower by using the modified D-A equation.

6.6 Fabric Filters

All three types of particulate removal devices mentioned earlier (gravity settling chambers, cyclones, and ESPs) share something in common. In these devices, a force is imposed on a particle in the gas stream to move it out of the gas streamline to hit a wall and get retained by the wall. The forces are gravitational force for gravity settling chambers, centrifugal force for cyclones, and electrostatic force for ESPs. The last two types of particulate removal devices in this chapter (i.e., fabric filters and wet scrubbers) work in a different way. In these devices the gas streams are divided by the collection targets (fabric media for filters and water droplets for wet scrubbers) and the particles are retained by the collection targets as the gas stream flows through the openings of the filter media or around the water droplets.

6.6.1 Principles of filtration

A fabric filter system consists essentially of a housing unit that host lots of fabrics to provide a large surface area for the gas stream to go through. Most of the particles in the incoming gas stream are retained by the fabric and the dust cake is gradually formed by the particles collected on the fabric. As the gas stream moves through, the dust accumulated in the cake and the filter fabric become the collection targets. The particle capture by these collection targets can be attributable to several mechanisms, including inertial impaction, Brownian diffusion, and electrostatic attraction (moderate electrostatic charges, negative or positive, can build on the surfaces of the dust cake, fabric, and particles). Sieving of particles by the pores of the cake is another removal mechanism.

Inertial impaction is the dominant removal mechanism for particles having $D_p >{\sim}1\mu$ while Brownian diffusion is moderately effective for collecting sub-micron particles. Different from most of the other particle removal devices, there are multiple opportunities for particles to encounter the collection targets. The fractional efficiency curve of fabric filters (Figure 6.23) illustrates that they are effective even in the difficult-to-control range of 0.2 to 0.5μ. It should also be noted that fabric filters have a much wider applicable range of 0.1 to $1,000\mu$ than most of the other particulate control devices (EPA, 2012).

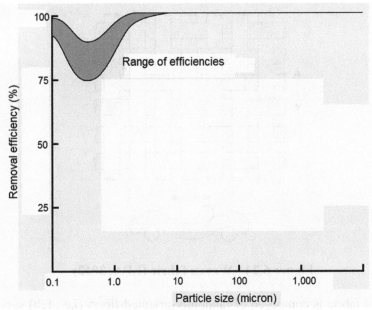

Figure 6.23 - Factional efficiency curve of fabric filters (modified from EPA, 2012)

6.6.2 Fabric filtration systems and their components

Two main functions of fabric material in a filtration system are to provide air passage and to support the filter cake. They need to tolerate temperature, chemical attack and physical abrasion. They may also need to be flex in some types of filter systems during the cleaning cycle. The two most common types of fabrics used are woven and felted. A woven fabric is composed of interlaced yarns at right angles to each other. The yarns in the "warp" direction provide the strength while those in the "fill" direction determine the characteristics of the fabric. The sizes of the pores can be >50µ (Figure 6.24).

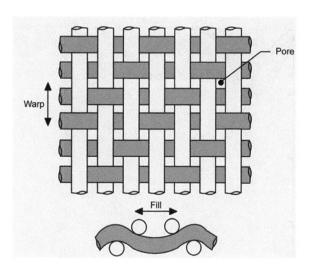

Figure 6.24 - Woven fabric (EPA, 2012)

A felted fabric is composed of randomly-oriented fibers (i.e., felt) supported by a very loose weave. The felted fabrics are usually thicker than the woven ones because there is a layer of felt on both sides of the weave support (Figure 6.25). The fibers on the dirty air side along with the dust cake provide numerous targets for incoming particles. In addition to woven and felted fabrics, membrane fabric, sintered metal fibers, and ceramic cartridges are also used (consult EPA (2012) for details).

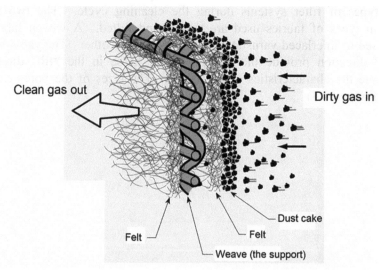

Figure 6.25 - Felted fabric (modified from EPA, 2012)

As the cake building up on the surface of the fabrics by more dusts removed from the gas stream, the head-loss (pressure drop) across the filter will go up accordingly. To avoid excessive pressure drops that would challenge the blower to deliver a specific gas stream flow rate and incur more energy costs, the filter material needs to be cleaned periodically. Based on the methods of cleaning, the commonly-used filters are grouped into (i) shaker, (ii) reverse-air, and (iii) pulse-jet fabric filters.

Figure 6.26 shows schematic of a *shaker fabric filter*. The open bottoms of the filter bags are attached to a tube sheet which provides a seal to separate the upper portion of the bags from the dust hopper. The closed tops of the bags are attached to a shaker mechanism. The particle-laden gas stream enters from the hopper, where some larger particles may settle out. The remaining particles will move with the gas stream upward and many of them will deposit on the inside of the bags as the gas stream moves through the bag. During the cleaning cycle, the dirty gas inlet to the fabric filter is closed. The bags are then mechanically shaken to dislodge the dust cakes into the collection hopper (EPA, 2012).

Construction of *reverse-air fabric filters* is very similar to that of the shaker fabric filters, except the closed tops are attached to a support structure, instead of a shaker mechanism. For the service cycle, the operations of these two types of filters are essentially the same. The cleaning cycle starts with stopping the influent of the dirty gas stream and closing the outlet of the filter. These filter systems usually have several compartments operated in parallel. Part of the clean gas stream effluent(s) from the other compartment(s) in service is directed to pass from the outside to the inside of the bags to dislodge the cakes and then moves through the dirty air inlet. Since the flow path of the gas in the cleaning cycle is opposite to that in the service cycle, that is how the name of "reverse-air" came from (EPA, 2012).

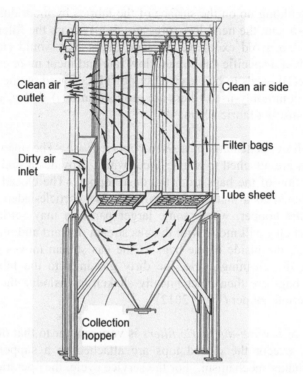

Clean air outlet

Clean air side

Filter bags

Dirty air inlet

Tube sheet

Collection hopper

Figure 6.26 - Shaker fabric filter
(Modified from EPA, 2012)

Figure 6.27 shows a *pulse-jet fabric filter* in which the tube sheet is located close to the top the filter unit to hold the tops of the bags. The gas stream enters the unit from the side of the casing or from the hopper, flows into the bags and leave the dust cake outside the bags, before it moves upward into the discharge outlet. The bags are also supported by metal cages inside to prevent them from collapsing. The cleaning cycle can be done concurrently within the service cycle on a row-by-row basis. A high-pressure pulse of compressed air is introduced at the top of each bag which will generate a pressure wave travelling downward inside the bag. The pressure wave will also induce some filtered gas flowing downward. The combined action will force the bags expand outward to crack the dust cake on the outside of the bags for the fall into the collection hopper. Off-line cleaning is more common and it is for filter systems with multiple compartments (EPA, 2012).

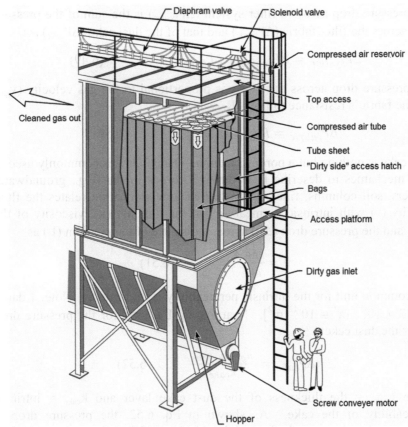

Diaphram valve
Solenoid valve
Compressed air reservoir
Top access
Compressed air tube
Tube sheet
"Dirty side" access hatch
Bags
Access platform
Dirty gas inlet
Screw conveyer motor
Hopper
Cleaned gas out

Figure 6.27 - Pulse-jet fabric filter
(Modified from EPA, 2012)

6.6.3 Factors affecting performance of fabric filters

Dust cake is the most important component of fabric filtration, governing the performance and operation of a given system. The initial particle removal efficiency of new filter bags is actually not as good before a sufficiently thick dust layer built on the fabric. The poorer performance may also appear immediately after the bags are placed back to service after cleaning. Excessive intensity, frequency, and duration of cleaning can increase particle emissions, at least for a short duration. On the other hand, a dust too thick may result in a larger pressure drop, which means a higher energy cost to push a specific gas flow rate through the filter.

The pressure drop across a filter system (ΔP_{system}) is the sum of the pressure drop across the filter fabric (ΔP_{fabric}) and that of the dust cake (ΔP_{cake}) as:

$$\Delta P_{system} = \Delta P_{fabric} + \Delta P_{cake} \qquad (6.49)$$

The pressure drop across the fabric is proportional to the gas velocity (v_{gas}) and the fabric's resistance to flow (R_{fabric}) as:

$$\Delta P_{fabric} = R_{fabric} \times v_{gas} \qquad (6.50)$$

Dust cake is nothing but a porous medium. Darcy's law is commonly used in fluid mechanics to describe flow through porous media (e.g., groundwater aquifers, soil columns, filter beds, etc.). Darcy's equation relates the fluid velocity (v) with intrinsic permeability of the medium (k), viscosity of the fluid, and the pressure drop (ΔP) across a distance of the medium (L) as:

$$v = \frac{k \Delta P}{\mu L} \qquad (6.51)$$

The common unit for the intrinsic permeability is m^2 or darcy [Note: 1 darcy $= 9.87 \times 10^{-13}$ $m^2 \cong 10^{-12}$ m^2]. Rearranging Eq. 6.51 for the pressure drop across the dust cake:

$$\Delta P_{cake} = \frac{\mu_g L_{cake} v_{gas}}{k_{cake}} \qquad (6.52)$$

where L_{cake} = the thickness of the dust cake layer and k_{cake} = intrinsic permeability of the cake. As shown in Eq. 6.52, the pressure drop is proportional to the gas velocity and the thickness of the cake.

The thickness of the dust cake will increase with time of the filter in service (Δt) and the mass loading of particulate loadings, which is proportional to v_{gas} and the particulate matter concentration in the gas stream (C_{PM}), thus

$$L_{cake} = (constant)(C_{PM})(v_{gas})(\Delta t) \qquad (6.53)$$

The constant in Eq. 6.53 accounts for unit conversions and the conversion of the retained mass into the volume of the dust cake. By inserting Eq. 6.53 to 6.52 and rearranging the above equations, the total pressure drop across the system becomes:

$$\Delta P_{system} = R_{fabric} v_{gas} + R_{cake} C_{PM} v_{gas}^2 \Delta t \qquad (6.54)$$

R_{cake} represents the resistance of the cake to flow. As shown in Eq. 6.54, the pressure drop across the dust cake increases with particulate concentration, the service time, and the square of the gas stream velocity.

The gas stream velocity (v_{gas}) is actually the ratio of the gas flow rate (Q_{gas}) and the surface area of the bags (A_{bag}) as

$$v_{gas} = \frac{Q_{gas}}{A_{bag}} \qquad (6.55)$$

This (Q/A) ratio in Equation 6.55 is often called the *air-to-cloth ratio, superficial velocity,* or *face velocity*. It is used to size the bags needed for a given gas flow rate. Using a larger air-to-cloth ratio to design a fabric filter, the required surface area would be smaller. However, as the air-to-cloth ratio increases, the *interstitial gas velocities* (the actual gas velocities) through the cake and the fabric increase. These higher velocities may push some of the smaller particles through these two barriers and emit from the system. In addition, the higher air-to-cloth ratios may adversely crush the existing cake structure so that uncovered openings on the fabric may happen to allow particles to go through.

For a filter system, there is a threshold air-to-cloth ratio. Below this threshold value, the particulate emission rates are relatively constant. Above this threshold, the emissions can increase sharply (Figure 6.28). The shaker fabric filters often use woven fabrics and operate with an air-to-cloth ratio of 2 to 4 ft/min (0.6 to 1.2 m/min). Reverse-air collectors usually use woven fabrics (membrane bags and felted bags are used in some applications) and operate with an air-to-cloth ratio of 1.5 to 3.5 ft/min (0.45 to 1.1 m/min). Pulse jet collectors use felted fabrics and operate with an air-to-cloth ratio of 3 to 10 ft/min (0.9 to 3.1 m/min) (EPA, 2012).

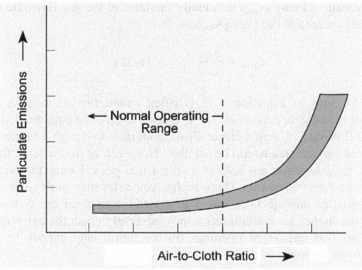

**Figure 6.28 - Effect of air-to-cloth ratio
(Modified from EPA, 2012)**

Example 6.20 Pressure drop across a dust cake

If the face velocity (i.e., the air-to-cloth ratio) to a fabric filter is 1 m/min, the thickness of the filter cake is 2 mm, and the intrinsic permeability of the cake is 1 darcy, estimate the pressure drop across the cake (T = 20 °C).

Solution:

Using Eq. 6.52:

$$\Delta P_{cake} = \frac{\mu_g L_{cake} v_{gas}}{k_{cake}} = \frac{(1.81 \times 10^{-5})(0.002)(1/60)}{(9.87 \times 10^{-13})} = 611 \frac{N}{m^2}$$
$$= 2.45 \; in - H_2O$$

6.6.4 Design calculations for fabric filters

As mentioned in the previous section, the air-to-cloth ratio can be used to estimate the required surface area of a fabric filter system; typical ranges were also given. More suggested air-to-cloth ratios can be found in literature. These general values should only be used as a starting point. A safety factor should always be used, especially when the dust loading is higher or the particle sizes are small. In addition, the calculated surface area is for the bags in service; the surface area of the bags, if they are off-line for cleaning or regular maintenance, is not part of the calculated value (EPA, 2012).

Example 6.21   Surface area of fabric filters

A shaker fabric filter has six compartments. During normal operation, five compartments are in service while the other one is off-line for cake removal or maintenance. Each compartment contains 100 bags that are 0.4 m in diameter and 6-meter long. The air-to-cloth ratio is 1.2 m/min.

(a) Calculate the available surface area of each bag.

(b) What is the air flow rate treated by this baghouse?

Solution:

(a) $A = \pi Dh = \pi(0.4)(6) = 7.54 \text{ m}^2$

(b) Number of bags in service = (5)(100) = 500

Surface area of bags in service = (500)(7.54) = 3,770 m^2

From Eq. 6.55,

$$\frac{Q}{A} = 1.2 \ m/min = \frac{Q_{gas}}{3,770 \ m^2}$$

$Q_{gas} = (1.2)(3,770) = 4,520 \text{ m}^3/\text{min}$

6.7 Wet Scrubbers

Wet scrubbers use a liquid stream, typically water, to collect particulates and/or gaseous pollutants from a gas stream. Wet scrubbers are often the control device of choice if the gas stream to be treated contains explosive gases or vapors. Chemicals can also be added to the scrubbing fluid to adjust certain characteristics of the gas stream (e.g., corrosiveness). It can also be a good alternative if simultaneous removal of particulates and gaseous pollutants is desirable.

6.7.1 Principles of wet scrubbing

Although wet scrubbers can collect both particles and gases at the same time, but usually they do not have high enough efficiency to perform both tasks simultaneously. This is mainly because they are designed to employ different removal mechanisms for particulates and gaseous pollutants. For effective removal of gaseous pollutants, wet scrubbers need to provide large air-water interfacial areas and relatively long residence time to facilitate the absorption of gaseous pollutants into liquid droplets (more details in the next chapter).

The main removal mechanisms involved in collecting particulate matter by wet scrubbers is impaction in which water droplets serve as the collection targets. Due to the limited residence time in most wet scrubbers, Brownian motion is typically not a significant removal mechanism. Some wet scrubbers use enhancements to improve their removal of fine particles by electrostatic charging and promotion of condensation (EPA, 2012).

There are many types of wet scrubbers and many of them have limited efficiencies for particles in the difficult-to-control range of 0.1 to 1.0 μ. The removal efficiency in this size range depends primarily on the intensity of the contact between the gas (and the particles) and the water in the scrubber. Some types of scrubbers (e.g., Venturi scrubbers) use high energies to develop large relative velocities between the particles and the collection targets (i.e., water droplets) to develop excellent inertial impaction efficiencies for particles in this size range. Figure 6-29 is a typical fractional efficiency curve that illustrates the range of performance for the various types of wet scrubbers.

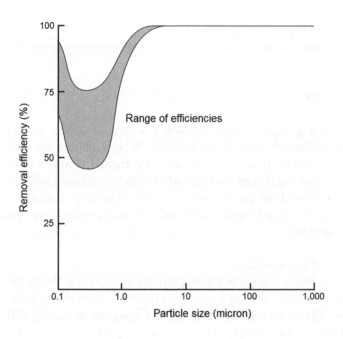

Figure 6.29 - Factional efficiency curve of wet scrubbers
(Modified from EPA, 2012)

6.7.2 Wet scrubbing systems and their components
There are different types of scrubber design to bring the contact between liquid droplets and particulates in the gas stream. Since inertial compaction is the main removal mechanism in scrubbers for particulate removal, the capability of a scrubber can be approximated from the pressure drop of the gas stream across the scrubber. A higher pressure drop implies more aggressive contact between the liquid and the gas streams, resulting a better removal of fine particles.

Scrubbers with static pressure drops of < ~5 in-H_2O (1.25 kPa) are referred to as *low-energy wet scrubbers*. They are, in general, capable of removing particles having diameters > 5 to 10 μ. *Medium-energy wet scrubbers* are capable of removing particles with sizes in microns and have pressure drops of 5 to 25 in-H_2O (1.25 to 6.25 kPa). Sub-micron particles can be effectively removed by high-energy wet scrubbers, but the pressure drop can be greater than 100 in-H_2O (25 kPa), depending on the particle size to be removed (EPA, 2012).

Spray tower scrubbers consist of a chamber, cylindrical or rectangular, with an array of spray nozzles mounted on multiple heads that are often spaced about one meter apart (Figure 6.30). Water droplets typically have a mean size of several hundred microns. They are an example of low-energy wet scrubbers; and it is simple and inexpensive. Because the relatively low velocity between the particles and the water droplets, they are only effective for particles larger than 5μ. They can be alternatives when control of both particulate matter and gaseous pollutants are needed (EPA, 2012).

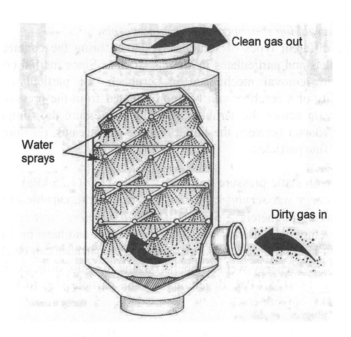

Figure 6.30 - Schematic of a spray tower scrubber
(Modified from EPA, 2012)

Packed-bed wet scrubbers are medium-energy wet scrubbers. They consist of a chamber containing a layer of packing layer. The packing is held in placed by wire-mesh retainer and supported by a plate. Scrubbing liquid is evenly distributed above the packing and moves downward through the packing bed to form a liquid film on the packing. The gas stream is forced to move upward through the packing to get in contact with the liquid (Figure 6.31). Packing materials come with different sizes and materials; some common types are shown in Figure 6.32. Although ceramic and metal packings are available, plastic packings are used in most of the air pollution applications (EPA, 2012).

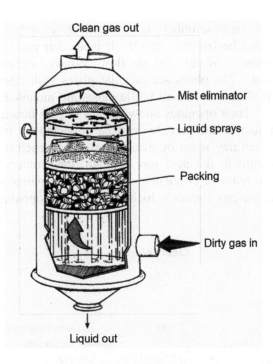

**Figure 6.31 - Schematic of a packed-bed scrubber
(Modified from EPA, 2012)**

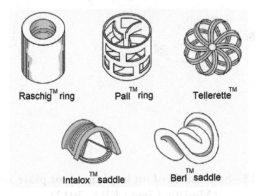

Figure 6.32 - Common types of packing (Modified from EPA, 2012)

If the upward air velocity is too strong through the packing bed, the water may not be able to travel downward. This limitation will result in reduction in removal efficiency. Removal efficiencies for fine particles ($< 3\mu$) are very low. Too high a particulate concentration may cause plugging in the packing materials (EPA, 2012).

An impingement-plate scrubber is a vertical chamber with several plates/trays mounted horizontally inside. It is another example of medium-energy wet scrubber. Water and air flow counter-currently, with water flowing downward. The plates are metallic plates with openings of ~3/16 inches (5 mm) in diameter. Small baffle plates are mounted directly above each of the holes. Their openings allow the gas to pass through and create a bubbling action so that the particles can be collected by impaction. The liquid depth on each tray is set by overflow weirs on each tray. The liquid overflows downward to the next tray through downcomers (Figure 6.33). They have limited removal efficiencies for particles smaller than 1μ, due to the limitation of the gas stream velocity through the openings of the trays (EPA, 2012).

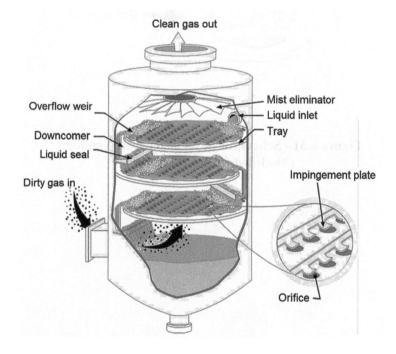

Figure 6.33 - Schematic of an impingement plate scrubber (Modified from EPA, 2012)

The Venturi scrubber is an example of a high-energy wet scrubber. It accelerates the gas stream to atomize the scrubbing liquid to enhance the gas-liquid contact. The fixed-throat Venturi scrubber (Figure 6.34) is one of the most common designs. The gas stream entering the converging section of the Venturi is accelerated to a velocity between 200 to 600 ft/s (60 to 180 m/s) at the inlet of the throat. The liquid is injected into the throat and becomes atomized into fine droplets. The size of the droplets depends on the throat gas velocity and the liquid-to-gas ratio; and typical mean size is 50 to 75 microns. The effectiveness of a Venturi scrubber depends on the relative velocity between the gas stream and the liquid droplets in the throat. They have the capability to remove sub-micron particles.

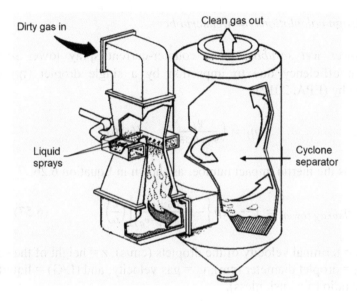

Figure 6.34 - Schematic of venturi scrubber
(Modified from EPA, 2012)

There are many other types of wet scrubbers available. One can consult EPA (2012) or EPA's Air Pollution Control Technology Fact Sheets.

6.7.3 Factors affecting performance of wet scrubbers
For any wet scrubber, the liquid flow rate is important to its removal performance. The rate of liquid flow is often expressed in terms of the liquid-to-gas ratio. Most wet scrubber systems for particulate removal operate with liquid-to-gas ratios between 4 to 20 gal/1,000 actual cubic feet; that is equivalent to a dimensionless ratio of about 0.0005 to 0.0025. Higher ratios do not usually enhance performance. Instead, they may exert a slight adverse impact due to changes in size distribution of the droplets. On the other hand, low liquid-to-gas ratios can have a severe adverse impact because of insufficient collection targets present.

Although a higher relative gas velocity across the packing bed, the tray openings, or Venturi throat will enhance its performance, it will incur higher power costs. In addition, it may create flooding in packing bed. The size of water drops in the spray scrubber is also an important design/operational parameter.

6.7.4 Design calculations for wet scrubbers

Spray tower wet scrubbers. For counter-current spray tower scrubbers, collection efficiency due to impaction by a single droplet (η_I) can be estimated by (EPA, 2012):

$$\eta_I = \left(\frac{\Psi_I}{\Psi_I + 0.35}\right)^2 \quad (6.56)$$

where Ψ_I is the inertia impact number as shown in Equation 6.26.

$$\eta_{spray\ tower} = 1 - exp\left\{-\left[\frac{1.5 u_t \eta_I z}{D_d(u_t - v_g)}\right]\left(\frac{L}{G}\right)\right\} \quad (6.57)$$

where u_t = terminal velocity of the droplets (cm/s), z = height of the scrubber (cm), D_d = droplet diameter (cm), v_g = gas velocity, and (L/G) = liquid to gas flow rate ratio (dimensionless).

Example 6.21 Performance of a spray tower scrubber
Estimate the removal efficiency of particles having D_p = 4µ (ρ_p = 1,200 kg/m³) for a gas stream (150 m³/min) by a counter-current spray tower (height = 3 m, T = 293K). the liquid flow rate is 120 L/min (diameter of the droplet = 500µ) and the gas velocity inside the scrubber is 1 m/s.

Solution:

(a) Use Eq. 6.12 to find the terminal velocity of the water droplet:

$$u_t = (1.0)\left[\frac{(9.81)(1,000 - 1.21)(500 \times 10^{-6})^2}{(18)(1.81 \times 10^{-5})}\right] = 7.52\ m/s$$

(b) Use Eq. 6.17 to find the Cunningham slip correction factor (C_c) for the 4µ particles:

$$C_c = 1 + \frac{6.21 \times 10^{-4}T}{D_p} = 1 + \frac{(6.21 \times 10^{-4})(293)}{4} = 1.05$$

(c) Use Eq. 6.26 to find the inertial impact number (Ψ_I):

$$\Psi_I = \left[\frac{(1.05)(4 \times 10^{-6})^2(1{,}200)(1.88 - 1.0)}{18(1.81 \times 10^{-5})}\right] \div (500 \times 10^{-6}) = 0.109$$

(d) Use Eq. 6.56 to find the removal efficiency of a single droplet (η_l):

$$\eta_l = \left(\frac{\Psi_I}{\Psi_I + 0.35}\right)^2 = \left(\frac{0.109}{0.109 + 0.35}\right)^2 = 0.056$$

(e) Use Eq. 6.57 to find the removal efficiency by the scrubber:

$$\eta_{spray\ tower} = 1 - exp\left\{-\left[\frac{1.5(7.52)(0.056)(3)}{(500 \times 10^{-6})(7.52 - 1.0)}\right]\left(\frac{0.12}{150}\right)\right\} = 37.2\%$$

Discussion:

1. In part (a), a value of one was used for the Cunningham slip factor because the droplet size is much larger than a few microns; and 1,000 kg/m^3 (the density of water) was used as the density of the droplet.

2. In part (b), the C_c for the 4μ particles was found to be 1.05, which is very close to unity.

3. In part (c), the relative velocity (the difference between the droplet velocity and the gas velocity) should be used (i.e., $7.52 - 1.0 = 6.52$ m/s).

4. The removal efficiency of 4-μ particles by this spray scrubber is relatively low, at 37.2%.

Packed-bed wet scrubbers. For packed-bed scrubbers, the removal efficiency can be estimated by:

$$\eta_{packed\ bed} = 1 - exp\left\{-\left[\frac{\pi z \Psi_I}{D_c \varepsilon (j - j^2)}\right]\right\} \qquad (6.58)$$

where z = scrubber height (cm), j = channel width as a fraction of packing diameter (dimensionless), ε = bed porosity, and D_c = packing diameter (cm).

The packing diameter, instead of water droplet diameter, should be used in calculating the inertial impact number. The channel width as a fraction of packing diameter (j) depends on the type of the packing used, but typically ranges from 0.165 to 0.0192. The bed porosity (ε) also depends on the type of packing used and typically ranges from 0.57 to 0.94. The values of these two parameters should be obtained from the manufacturers.

Example 6.22 Performance of a packed-bed scrubber

Referring to Example 6.21, a packed bed is to be used instead (packing = 5 cm Raschig^{TM} rings). Assume j = 0.165 and porosity = 0.75.

Solution:

(a) Use Eq. 6.26 to find the inertial impact number (Ψ_1):

$$\Psi_I = \left[\frac{(1.05)(4 \times 10^{-6})^2(1,200)(1.0)}{18(1.81 \times 10^{-5})}\right] \div (0.05) = 1.24 \times 10^{-3}$$

(b) Use Eq. 6.58 to find the removal efficiency by the packed-bed scrubber:

$$\eta_{packed\ bed} = 1 - exp\left\{-\left[\frac{\pi(3)(1.24 \times 10^{-3})}{(0.05)(0.75)(0.165 + 0.165^2)}\right]\right\} = 0.802$$
$$= 80.2\%$$

Discussion: In part (a), the target diameter is the size of the packing (0.05 m) and the velocity is the gas velocity (1 m/s).

Impingent-plate wet scrubbers. For tray scrubbers, the removal efficiency can be estimated by:

$$\eta_{tray} = 1 - [exp(-80F^2\Psi_I)]^n \quad (6.59)$$

where F = foam density function (dimensionless) and n = number of trays.

The gas velocity through the holes in the plate and the diameter of the holes, instead of the diameter of the water drops should be used to calculate the inertial impact number. The foam density fraction typically ranges from 0.38 to 0.65.

Example 6.23 Performance of a tray scrubber

Referring to Example 6.21, a tray scrubber is to be used instead (3 trays with 10 mm holes; gas velocity through the holes is 18.0 m/s). Assume the foam density function (F) = 0.50.

Solution:

(a) Use Eq. 6.26 to find the inertial impact number (Ψ_1):

$$\Psi_I = \left[\frac{(1.05)(4 \times 10^{-6})^2(1,200)(18.0)}{18(1.81 \times 10^{-5})}\right] \div (0.01) = 0.106$$

(b) Use Eq. 6.58 to find the removal efficiency by the tray scrubber:

$$\eta_{tray} = 1 - \left[exp\left(-80(0.50)^2(0.106)\right)\right]^3 = 0.998 = 98.8\%$$

Discussion: In part (a), the target diameter is the size of the hole (0.01 m) and the velocity is the gas velocity through the hole (18.0 m/s), which is much larger than the gas velocity though the bulk of the scrubber (1 m/s).

Venturi scrubbers. For a Venturi scrubber, the liquid is atomized by the gas stream into droplets. The droplet diameter depends on the gas velocity through the throat ($v_{g,throat}$) and the liquid-to-gas ratio. The Sauter mean diameter is considered to be most representative as the diameter of the water droplet in venturi scrubbing. The *Sauter mean diameter* is the diameter of a droplet/particle having the same volume/area ratio as the entire size distribution. For a gas-liquid system, this droplet diameter can be estimated as (EPA, 2012):

$$D_{Sauter} = \frac{16,400}{v_{g,throat}} + 1.45 \left(\frac{L}{G}\right)^{1.5} \qquad (6.60)$$

where D_{Sauter} = mean droplet diameter (μ), $v_{g,throat}$ = gas velocity at the throat (ft/sec), and (L/G) = the liquid to gas ratio (gal/1,000 ft^3).

For venturi scrubbers, the removal efficiency can be estimated by:

$$\eta_{venturi} = 1 - exp\left\{-k\sqrt{\Psi}\left(\frac{L}{G}\right)\right\} \qquad (6.61)$$

where k = an empirical constant, typically 0.1 to 0.2 (in 1,000 ft^3/gal) and (L/G) = the liquid to gas ratio (in gal/1,000 ft^3)

Equations 6.60 and 6.61 are in the US customary units. For use of SI units, they can be modified to:

$$D_{Sauter} = \frac{5,000}{v_{g,throat}} + 646,900 \left(\frac{L}{G}\right)^{1.5} \qquad (6.62)$$

where D_{Sauter} = mean droplet diameter (μ), $v_{g,throat}$ = gas velocity at the throat (m/sec), and (L/G) = the liquid to gas ratio (dimensionless).

$$\eta_{venturi} = 1 - exp\left\{-k\sqrt{\Psi}\left(\frac{L}{G}\right)\right\} \qquad (6.63)$$

where k = an empirical constant, typically 750 to 1,500 and (L/G) = the liquid to gas ratio (dimensionless).

Example 6.24   Performance of a Venturi scrubber (the US customary units)

Estimate the removal efficiency of particles having $D_p = 1\mu$ ($\rho_P = 1,500$ kg/m³) by a venturi scrubber having a throat velocity of 328 ft/s (100 m/s) and a liquid-to-gas ratio of 8.0 gallon/1,000 ft³ (= 1.07×10^{-3} dimensionless). Assume T = 68 °F (20 °C) and k = 0.15 1000 ft³/gal.

Solution:

(a) Use Eq. 6.60 to calculate the mean droplet diameter (D_{Sauter}):

$$D_{Sauter} = \frac{16,400}{328} + 1.45(8.0)^{1.5} = 82.8\mu$$

(b) Use Eq. 6.17 to find the Cunningham slip correction factor (C_c) for the 1μ particles:

$$C_c = 1 + \frac{6.21 \times 10^{-4}T}{D_p} = 1 + \frac{(6.21 \times 10^{-4})(293)}{1} = 1.18$$

(c) Use Eq. 6.26 to calculate the inertial impact number (Ψ_1):

$$\Psi_I = \left[\frac{(1.18)(1 \times 10^{-6})^2(1,500)(100.0)}{18(1.81 \times 10^{-5})}\right] \div (82.8 \times 10^{-6}) = 6.56$$

(d) Use Eq. 6.61 to find the removal efficiency by the Venturi scrubber:

$$\eta_{venturi} = 1 - exp\{-0.15\sqrt{6.56}(8.0)\} = 0.954 = 95.4\%$$

Discussion: In part (b), the target diameter is the mean size of the liquid droplets (82.8μ).

Example 6.25   Performance of a Venturi scrubber (the SI units)

Referring to Example 6.24, use the values in SI units and Equations 6.22 and 6.23 to estimate the removal efficiency of particles having $D_p = 1\mu$. The corresponding dimensionless k value for k of 0.15 1000 ft³/gal is 1,122 (= 0.15×7,480).

Solution:

(a) Use Eq. 6.60 to calculate the mean droplet diameter (D_{Sauter}):

$$D_{Sauter} = \frac{5,000}{100} + (938,000)(1.07 \times 10^{-3})^{1.5} = 82.8\mu$$

(b) Use Eq. 6.26 to calculate the inertial impact number (Ψ_I):

$$\Psi_I = \left[\frac{(1.18)(1 \times 10^{-6})^2(1,500)(100.0)}{18(1.81 \times 10^{-5})}\right] \div (82.8 \times 10^{-6}) = 6.56$$

(c) Use Eq. 6.63 to find the removal efficiency by the Venturi scrubber:

$$\eta_{venturi} = 1 - exp\{-(1,122)\sqrt{6.56}(1.07 \times 10^{-3})\} = 0.954 = 95.4\%$$

Discussion: $1,000 \text{ ft}^3 = 7,480$ gallons.

6.8 Considerations in Selection of Types of Removal Devices

Selecting a device to remove/collect particles from a gas stream, characteristics of the particles (e.g., PSD, solid/liquid, stickiness, and greasiness), characteristics of the gas stream (e.g., temperature, humidity, and presence of acidic gas), gas stream flow rate, mass concentrations of the pollutants, and size of the particles to be collected need to be taken into consideration.

Below are some general suggestions:
1. Post-treatment of the collected dusts/liquids should be considered as part of the treatment process.
2. Temperatures of gas stream may be needed. Tolerance of construction materials to high temperatures should be considered. For example, materials in fabric filters should have proper temperature rating.
3. Reducing the temperature of the gas stream would reduce its flow rate. This may be beneficial because it will reduce the pressure drop across the device (less energy cost) as well as increase the residence time in a control device (better treatment performance). However, it may cause condensation which can be troublesome, if acidic gas components are present.
4. Sticky and greasy particles may create problems on collection surfaces of the cyclone, fabric filter, or ESP.
5. Wet scrubbers would generally be a better choice for explosive gases than the other types of control devices.

6. Cut size of gravity settling chambers is about 50μ and that of cyclones is about 5μ. If fine particles also need to be removed, gravity settling chambers and/or cyclones can serve as pretreatment for fabric filters, ESP, or wet scrubbers which can remove finer particles.

7. Particles in the gas stream may be abrasive to the control devices.

6.9 Conclusion

1. Sizes of particles in gas streams are typically log-normally distributed. Fine particles, having diameters <10 μ, are of greater concern with regards to air pollution.

2. Major particle removal mechanisms in common control devices are gravitation, centrifugation, electrostatic attraction, inertial impaction, and Brownian motion.

3. Common particulate removal devices include gravity settling chambers, cyclones, electrostatic precipitators, fabric filters, and wet scrubbers.

4. In gravity settling chambers, cyclones, and ESPs, particles are driven a collection surface, mainly by gravitational, centrifugal, and electrostatic forces, respectively. The design equations are relatively similar.

5. In fabric filters and wet scrubbers, the gas stream is divided by the collection targets (i.e., fabric in the filters and water droplets in the wet scrubbers). The particulates will be collected by the collection targets, while the gas stream moves around them. The design equations are different.

6. Cut size of gravity settling chambers is about 50μ and that of cyclones is about 5μ. If fine particles also need to be removed, gravity settling chambers and/or cyclones can serve as pretreatment to fabric filters, ESP, or wet scrubbers which can remove finer particles.

Bibliography

Calvert, S.; Goldschmid, J.; Leith, D.; and Mehta, D (1972). *Scrubber Handbook. Vol. I, Wet Scrubber System Study.* EPA-R2-72-118a, U.S. Environmental Protection Agency.

de Nevers, N (2000). *Air Pollution Control Engineering (2nd edition)*, McGraw-Hill Companies, Inc.

Engineering Science (1979). *Scrubber Emissions Correlation* Final Report. U.S. EPA Contract 68-01-4146, Task Order 49, U.S. Environmental Protection Agency.

Kuo, J. (2014). *Practical Design Calculations for Groundwater and Soil Remediation* (2nd edition), CRC Press, Boca Raton, Florida.

Kuo, J.F. and Cordery, S.A. (1988). *Discussion of Monograph for Air Stripping of VOC from Water*, J. Environ. Eng., V. 114, No. 5, p. 1248-50.

LADCO (2009). *APTI 400: Introduction to Air Toxics – Student Manual.* Lake Michigan Air Directors Consortium (LADCO), Rosemont, Illinois 60018.

USEPA (1981). *APTI SI 422: Air Pollution Control Orientation Course – Unit 3 Air Pollution Meteorology (3rd edition)*, EPA 450/2-81-017C, prepared by Northrop Services, Inc. for United States Environmental Protection Agency, Research Triangle Park, NC 27711.

USEPA (1982). *Control Techniques for Particulate Emissions from Stationary Sources. Vol. 1, Wet Scrubbers.* EPA 450/3-81-005a, U.S. Environmental Protection Agency.

USEPA (1999). *APTI 444: Air Pollution Field Enforcement - Student Manual*, Office of Air Quality Planning and Standards, United States Environmental Protection Agency, Research Triangle Park, NC 27711.

USEPA (2002). *EPA Air Pollution Control Cost Manual (6th edition).* EPA/452/B-02-001. Office of Air Quality Planning and Standards, United States Environmental Protection Agency, Research Triangle Park, NC 27711.

USEPA (2003). *APTI 445: Inspection of Particle Control Devices - Student Manual (2nd revision)*, Air Pollution Training Institute (APTI), Environmental Research Center, United States Environmental Protection Agency, Research Triangle Park, NC 27711.

USEPA (2003). *APTI 452: Principles and Practices of Air Pollution Control - Student Manual (3rd edition)*, Air Pollution Training Institute, United States Environmental Protection Agency, Research Triangle Park, NC 27711.

USEPA (2012). *APTI 413: Control of Particulate Matter Emissions.* Air Pollution Training Institute (APTI), Environmental Research Center, United States Environmental Protection Agency, Research Triangle Park, NC 27711.

USEPA (2012). *APTI 413: Control of Particulate Matter Emissions - Student Manual (5th Edition)*, prepared by Tidewater Operations Center of C^2 Technologies for Air Pollution Training Institute, United States Environmental Protection Agency, Research Triangle Park, NC 27711.

Exercise Questions

1. To determine the diameter of a small particle (density = 2.5 g/cm^3), we let it settle in still air (T = 20 °C) in the field of view of a microscope. The settling/terminal velocity was 0.005 m/s. Use the Stokes' law to calculate the particle diameter (in microns).

2. The separation number for a 2-μ particle (density = 2.0 g/cm^3) in an air stream against a sphere was calculated to be 1.0. Estimate the separation number for a 1-μ particle (density = 2.5 g/cm^3) in the same air stream.

3. To determine the diameter of a small particle (density = 2.5 g/cm^3), we let it settle in still air (T = 20 °C) in the field of view of a microscope. The settling/terminal velocity was 0.005 m/s and the particle diameter was calculated to be 8μ. What is the Reynolds number of this particle travelling at the stated velocity?

4. A spherical particle with a diameter of 5μ and density of 3,000 kg/m^3 is settling in the still air (T = 20 °C; P = 1 atm; viscosity = 1.81 × 10^{-5} kg/m•s) and the Cunningham correction factor = 1.0.

    (a) Determine its terminal velocity in this air (in m/s)

    (b) Determine its Reynolds number at this settling speed.

    (c) What is its aerodynamic diameter?

    (d) What would be its Stokes stopping distance if it is ejected at 10 m/s?

    (e) Determine the separation number if this particle is moving at 5 m/s against a stationary 1-cm sphere.

5. A spherical particle with a diameter of 2μ and density of 2,000 kg/m^3 is settling in the still air (T = 20 °C and P = 1atm). Its terminal velocity is 2.4 × 10^{-4} m/s; Reynolds number at this velocity is 3.2 ×10^{-5} and diffusivity in this air is 1.2 ×10^{-11} m^2/s. For a spherical particle with a diameter of 4μ and density of 3,000 kg/m^3 in the same air, determine

    (a) its terminal velocity in this air (in m/s)

    (b) its Reynolds number at the settling speed

    (c) its diffusivity in this air (in m^2/s).

6. A log-normally distributed particle mass has a geometric mean diameter of 10μ and the diameter of particles with 84.13% by weight of all particles in this particle mass smaller is 25μ. Determine

    (a) the geometric standard deviation (in μ)

(b) the standard deviation (in μ)

(c) the diameter of particles with 15.87% by weight of all particles smaller (in μ).

7. A cyclone separator has an inlet width of 0.5 meter treating an air stream (viscosity $= 2.1 \times 10^{-5}$ kg/m•s) with an inlet gas velocity of 14 m/s and number of turns within the cyclone is 5. The density of all particles is 1,500 kg/m^3. Determine

(a) the cut diameter of this cyclone, using the block flow model (in μ)

(b) the removal efficiency for particles of 10μ (in %).

8. A cyclone separator was operated at 500 scfm with $D_{cut} = 5\mu$. However, the flow has recently been decreased to 250 scfm. Without any modifications to this cyclone,
(a) what is the new cut size?
(b) will the cyclone provide higher removal efficiencies at this reduced flow rate?

9. A cyclone separator was operated at 800 scfm with $D_{cut} = 5\mu$. However, the flow has been increased to 1,000 scfm due to the expansion in manufacturing activities.

(a) Without any modification to this cyclone, what is the new cut size?

(b) What is the new removal efficiency for 5-μ particles?

10. A gas stream contains particles of two sizes: 20 μ (30% by weight) and 2μ (70% by weight). The air is to be treated by a cyclone ($D_{cut} = 4\mu$). Estimate the collection efficiency (% by weight) of all particles?

11. A cyclone has an outside diameter of 1 m and a rectangular inlet (Height $= 0.4$ m and Width $= 0.2$ m). The gas flow rate is 2.4 m^3/s (85 ft^3/s) carrying particles with a density of 3,000 kg/m^3.

(a) Demonstrate the gas flow velocity inside the cyclone is equal to 30 m/s;

(b) Determine the cut diameter with N = 5 and viscosity of air = 1.8 $\times$ 10^{-5} kg/m•s;

(c) Estimate its removal efficiency for particles with diameter = 2μ

12. A cyclone separator is being operated at 1,000 cfm with $D_{cut} = 5\mu$. With an improvement in the production, the air stream flow rate will be reduced to 600 cfm.

(a) Without any modification to this cyclone, what is the new cut size?

(b) Will the reduction in the inlet flow rate improve the particle removal efficiency of the cyclone?

13. A gas stream contains particles of two sizes: 20 μ (60% by weight) and 2μ (40% by weight). The air is to be treated by a cyclone (D_{cut} = 5μ). Estimate the collection efficiency (% by weight) of all particles?

14. A bag house has 10 compartments (200 bags per compartment with bag length of 12 feet and diameter of 6 inches) and used for controlling particulate emission from an air stream (Q = 60,000 ft³/min). Calculate the air-to-cloth ratio (ft³/min/ft²) when two compartments are taken off-line for bag cleaning and maintenance.

15. A 500-MW coal-fired plant produces 1 million acfm of stack gas to be treated by a shaker bag house. From a design handbook, the engineer found the typical air/cloth ratio of 4 cfm/ft² is appropriate for this type of application. Estimate the number of bags (40 ft long and 1 ft in diameter) needed.

16. A bag house is to be constructed to control particulate emission from an air stream (Q = 30 m³/s). The filter bags to be use measure 0.5 m in diameter and 4.0 m in length. The engineer decided to use the air/cloth ratio (1 m³/min/m²) found from a design handbook for design. Determine the minimum number of bags needed for the bag house.

17. A shake-deflate baghouse have six compartments. During normal operation, five compartments are in service while the other one is off-line for cake removal or maintenance. Each compartment contains 100 bags that are 0.4 m in diameter and 6-meter long. The air-to-cloth ratio is 1.2 m/min.

(a) Calculate the available surface area of each bag

(b) What is the air flow rate treated by this baghouse?

18. Assume Deutsch-Andersen equation applies. An ESP is treating a particle-laden air stream (T = 600 °F), collecting 90% of the particles at an (A/Q) ratio of 0.2 min/ft.

(a) Determine the drift velocity of these particles?

(b) Assuming the drift velocity stays the same, if the air stream is cooled down to 300 °F before entering the ESP, will the removal efficiency increase or decrease?

(c) What would be the new removal efficiency for these particles?

19. An ESP system is designed to treat an air stream of 500 m³/min (T = 500 °C) with 90% particulate removal efficiency. However, the recent regulatory requirement requires a 98% removal. The owner decided to lower the temperature of the air to 200 °C. Will this approach meet the new requirement (assume drift velocity is a constant and use D-A equation)?

20. A gas stream contains particles of two sizes: 2 and 4 microns. The particle density is the same for both sizes. Assume the Deutsch-Anderson equation applies and the drift velocity is proportional to the diameter. If the collection efficiency for the 4-μ particles is 99%, estimate the collection efficiency for the 2-μ particles.

21. An ESP system is designed to treat an air stream of 500 m^3/min. The system will provide a 95% removal efficiency for particles with drift velocity of 0.2 m/s. The size of each plate is 5 meters tall $\times$ 2 meters long. Calculate the minimum number of plates needed using the Deutsch-Anderson equation.

22. An electrostatic precipitator is designed to treat an air stream of 500 m^3/min. The system is to remove particles with a drift velocity of 0.2 m/s. Using the Deutsch-Anderson equation to determine

(a) the plate area required for 90% removal efficiency (in m^2)

(b) the plate area required for 99% removal efficiency (in m^2)

(d) the plate area required for 99.9% removal efficiency (in m^2).

23. Estimate the removal efficiency of particles having $D_p = 5\mu$ ($\rho_p = 1{,}500$ kg/m^3) for a gas stream (160 m^3/min) by a counter-current spray tower (height = 4 m, T = 293K). The liquid flow rate is 150 L/min (diameter of the droplet = 400μ) and the gas velocity inside the scrubber is 1.2 m/s.

24. Referring to Question #23, a packed bed is to be used instead (packing = 5 cm Raschig^{TM} rings). Assume j = 0.165 and porosity = 0.75. Determine its removal efficiency.

25. Referring to Question #23, a tray scrubber is to be used instead (3 trays with 10 mm holes; gas velocity through the holes is 18.0 m/s). Assume the foam density function (F) = 0.50. Determine its removal efficiency.

26. Estimate the removal efficiency of particles having $D_p = 0.8\mu$ ($\rho_P = 2{,}000$ kg/m^3) by a Venturi scrubber having a throat velocity of 328 ft/s and a liquid-to-gas ratio of 9.0 gallon/1,000 ft^3. Assume T = 68 °F and k = 0.16 1000 ft^3/gal.

27. Estimate the removal efficiency of particles having $D_p = 0.8\mu$ ($\rho_P = 2{,}000$ kg/m^3) by a Venturi scrubber having a throat velocity of 100 m/s and a liquid-to-gas ratio of 0.0012 (dimensionless). Assume T = 20 °C and k = 1,200.

Chapter 7

Control of Organic and Inorganic Gaseous Emission

Gaseous pollutants in ambient air come from fugitive and point sources. Reducing emissions of gaseous air pollutants is vital to achieve good ambient air quality. Fugitive emission and control will be covered in Chapter 11. There are many types of control devices for removing gaseous air pollutants from stationary sources. Removal mechanisms of these devices vary. Characteristics of pollutants and of the gas stream will have significant impacts on their removal.

This chapter starts with an introduction on sources of gaseous points, types of common pollution reduction devices, and characteristics of gas streams that are relevant to pollutant removal (Section 7.1). Adsorption, absorption, thermal oxidation, catalytic incineration, condensation, and biodegradation are processes commonly used to reduce point-source emissions of gaseous pollutants. Section 7.2 describes principles, commonly-used systems and their components, important design considerations, and design calculations related to adsorption. Those of absorption, thermal oxidation, catalytic incineration, condensation, and biodegradation are presented in Sections 7.3, 7.4, 7.5, 7.6, and 7.7, respectively.

7.1 Introduction

Gaseous pollutants in ambient air can be primary or secondary. Unless for indoor air, it is uncommon for us to draw ambient air into a treatment unit to remove air pollutants. Instead, elimination and reduction of man-made emissions play an important role in achieving good ambient air quality. This chapter focuses on reductions of primary gaseous pollutants from point sources. These reductions are typically achieved by applying an end-of-the-pipe control system (i.e., add-on control devices).

Primary gaseous pollutants can be inorganic or organic. Organic gaseous pollutants include volatile organic compounds (VOCs) and other organic compounds. Inorganic gaseous pollutants of concern include SO_x, NO_x, CO, GHGs, acidic gases (e.g., HCl and HF), and H_2S and other reduced sulfur compounds. Emission reductions of SO_x, NO_x, CO, and GHGs are covered in separate chapters in this book. Many treatment processes for organic compounds (e.g., adsorption, absorption, thermal processes and biological processes) are applicable to acidic gases and reduced sulfur compounds. Consequently, this chapter focuses on end-of-the-pipe treatment of organic gaseous pollutants.

7.1.1 Sources and emissions of VOCs

Table 7.1 tabulates sources and emissions of VOCs in the U.S. in 2017. The total VOCs emission is about 16.2 million tons in 2017. Contributions to VOCs emissions are 46.6, 28.9, 21.3, and 3.2% from the industrial and other processes, miscellaneous, transportation and stationary fuel combustion sectors, respectively. Petroleum and related industries are the largest industrial sources and account for 19.4% of the total emission, while solvent utilization is the second largest industrial source (18.8% of the total emission).

Figures 7.1 depicts trends of annual VOCs emissions in the U.S. since 1970. As shown, the emissions have been decreasing, and most of the reductions come from the transportation sector and some from the industry sector.

Table 7.1 - VOCs emission rates and source categories in the U.S. in 2017

Source Category	(1,000 tons)	(%)
Stationary fuel combustion	**519**	**3.2**
Electric utility	38	0.2
Industral	110	0.7
Others	372	2.3
Industrial and other Processes	**7,557**	**46.6**
Chemical & allied product manufacturing	77	0.5
Metals processing	29	0.2
Petroleum & related industries	3,145	19.4
Other industrial processes	346	2.1
Solvent utilization	3,052	18.8
Storage & transport	675	4.2
Waste disposal & recycling	233	1.4
Transportation	**3,457**	**21.3**
Highway vehicles	1,801	11.1
Off-Highway	1,656	10.2
Miscellaneous	**4,699**	**28.9**
Wildfires	2,466	15.2
Others	2,232	13.8
Total	**16,232**	**100.0**

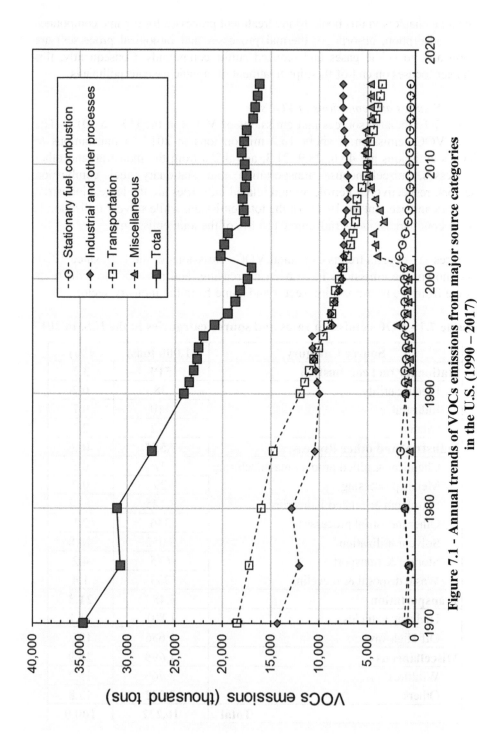

Figure 7.1 - Annual trends of VOCs emissions from major source categories in the U.S. (1990 – 2017)

7.1.2 Types of emission reduction technologies

End-of-the-pipe control processes for gaseous pollutants can be classified into two categories: recovery and destruction. Recovery processes refer to transferring compounds of concern (COCs) in a gas stream to collection media which can be solid or liquid. Common recovery processes include adsorption, absorption, and condensation. Destruction processes include thermal oxidation (direct or catalytic incineration), chemical oxidation, and biodegradation that convert COCs into reaction products that are more acceptable. Although chemical oxidation has reportedly been used for destructing gaseous pollutants, it is not commonly practiced in the field.

7.1.3 Important gas stream characteristics

As mentioned, many different technologies are available for emission reductions of gaseous pollutants. In addition to economical consideration, the selection and design of a treatment system will depend greatly on the characteristics of the gas stream. They include flow rate, temperature, pressure, pollutant concentrations, heating values, oxygen level, moisture content, and particulate concentrations.

Data on flow rate, temperature, and pressure are needed to size a treatment device to meet design operational conditions. They can also be used to determine the flow velocity through the treatment unit which is another important operating parameter. From the flow rate and the pollutant concentration, the mass loading to the control device, another important operating parameter, can be determined. Heating values are of concern for the thermal processes. Oxygen level and moisture content are parameters of concern for both thermal and biological processes. For adsorption, moisture content and particulates are of concern, because water molecules may compete with COCs for the limited adsorption sites and particulates may mask those sites.

7.2 Adsorption

Adsorption processes have been widely used for effective removal of a wide variety of gaseous organic and inorganic (e.g., H_2S) pollutants. Adsorption is a surface phenomenon. For air pollution control, adsorption is to remove undesirable COCs from a gas stream by having them attached/adhered onto the surface of a solid. The participants of this process include: *adsorbent* (the solid), *adsorbate* (the COCs) and the *carrier gas* (the inert portion of the gas stream, typically air).

6.2.1 Principles of adsorption

Adsorption is basically a mass transfer process from the gas to the solid. The COCs travel through the bulk gas phase to the adsorbent, enter into pores of the adsorbent (mainly due to diffusion caused by the concentration gradient), accumulate in the pores, and get adsorbed onto the surface of the adsorbent (Figure 7.2).

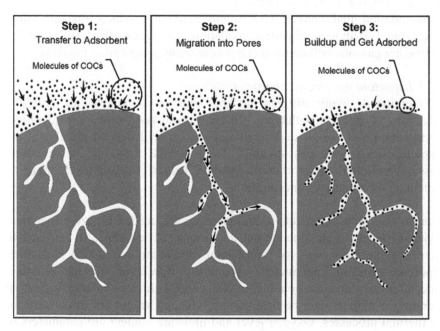

Figure 7.2 - Adsorption Steps (modified from EPA, 2012)

Adsorption can be a physical or chemical process. In *chemical adsorption*, a chemical bond is formed between the adsorbent and the COC molecule; while in *physical adsorption*, the COC molecule is retained by the solid surface through intermolecular cohesion only. In most industrial applications, the physical adsorption is the preferred mode since it is readily reversible (i.e., *desorption*) when the operational conditions change. Desorption is an approach to regenerate spent adsorbent; and *regeneration* is to refresh the adsorbent. It should be noted that the adsorption process is exothermic (heat will be released as a result of adsorption). Consequently, lower temperatures are more favorable for adsorption while higher temperatures are better for desorption.

6.2.2 Adsorption systems and their components

The main component of an adsorption system is the adsorbent. An effective adsorbent should have a large specific surface area with good affinity for COCs. The *specific surface area* can be defined as the total surface area per unit mass (m^2/g). Common types of adsorbents are activated carbon, zeolites, synthetic polymers, activated alumina, and silica gel.

Activated carbon is the most commonly-used adsorbent for air pollution control. It can be produced from different raw carbonaceous materials. The raw material first goes through pyrolysis to drive off all volatile materials and the remaining carbon is then "activated" by steam, air or CO_2 to have a porous and amorphous structure with a large specific surface area. The activated carbon can be in the form of granules or powders. For air pollution control, use of granular activated carbon (GAC) is more common. For example, the size of GAC for an application is 20 × 40 mesh (i.e., 90% or more of the GAC will pass through a 20-mesh sieve (0.810 mm opening) and retained by a 40-mesh sieve (0.420 mm)). Due to its non-polar surface, GAC is applicable for a wide variety of organic and inorganic gases. Typical internal porosity, specific surface area, dry bulk density, and mean pore diameter of GAC are 55 to 75%, 600 to 1,600 m^2/g, 0.35 to 0.50 g/cm^3, and 1,500 to 2,000 Å [Note: 1 Angstrom = 10^{-10} m], respectively (EPA, 2012).

Zeolites (molecular sieves) can be natural or synthetic with a crystalline structure. They have been used to remove moisture and NO_x from air pollution sources. Synthetic polymeric adsorbents have a variety functional groups and high adsorption capacities for selected organic compounds, such as ketones, aldehydes, and reactive compounds, that may have unwanted reactions on the surface of GAC. They are also used for gas streams with relative humidity (RH) greater than 50%, because they have less affinity to water vapor than GAC has. Silica gel and activated alumina are mainly used to remove moistures in a gas stream (EPA, 2012).

Each type of adsorbent has its own maximum capacity, or a limited number of active sites, for adsorption. Once the adsorbed COCs occupy most of the available sites, the removal efficiency will drop significantly. If the operation is continued beyond this point, the breakthrough point (defined later) will be reached and the effluent concentration will rise sharply. Eventually, the adsorbent in service would be saturated, exhausted, or spent, when most of its sites are occupied. Adsorbers have been employed for a wide range of COC concentrations. For odor control and low COC concentration applications (<10 ppm), the adsorbent bed is often discarded when the adsorbent approaches saturation with the COCs. These are non-regenerative systems because the adsorbent is not regenerated for reuse.

195

The non-regenerative adsorbers often consist of thin adsorbent beds (Figure 7.3(a) as an example), with thickness ranging from 1.3 to 10 cm (0.25 to 4 inches). The pressure drop across a bed is relatively low (< 2.5 in-H_2O, or 0.62 kPa) with a typical superficial gas velocity of 6 to 20 m/min (20 to 60 ft/min) through the bed. The *superficial gas velocity* can be determined by dividing the gas flow rate with the cross-sectional area of the adsorbent bed. Design service lives of these adsorbent beds range from six months for heavy odor concentrations and up to two years for trace concentrations or intermittent operations. Figure 7.3(b) illustrates a thick-bed non-regenerative adsorber which is essentially housed in a 55-gallon (0.2 m^3) drum (EPA, 2012).

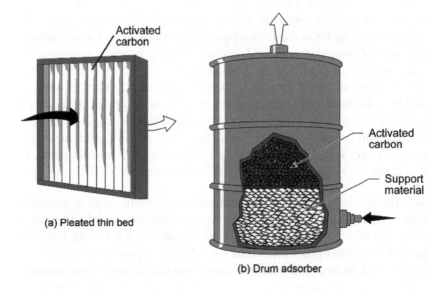

Figure 7.3 - Non-regenerative adsorbers: (a) pleated thin bed and (b) drum adsorber (Modified from EPA, 2012)

For applications with high COC concentrations (up to 10,000 ppm), it would become uneconomical to discard the exhausted adsorbent after it reaches its first saturation with the COCs. Instead, the adsorbent is regenerated to refresh for the next cycle and the previously-adsorbed COCs may also be recovered for potential recycle/reuse. Figure 7.4 illustrates a 3-bed fixed-bed regenerative adsorption system. In the operation mode, one or more beds treat the influent gas stream while the remaining one(s) are being regenerated or cooled down. The regeneration is often by low-pressure steam (or hot air, vacuum) in which the adsorbed COCs will be desorbed from the adsorbent and move into the steam. The mixture of the steam and the COCs is then chilled and condensed for separation into the COCs and water. As shown, there is a particulate filter to remove particles which may mask the active adsorption sites and a heat exchanger to lower the gas temperature to a range of 15 to 38 °C (60 to 100 °F) for optimal adsorption efficiency and adsorbent service life (EPA, 2012). In addition to the fixed-bed design, there are moving-bed and fluidized-bed design (see EPA (2012) for details).

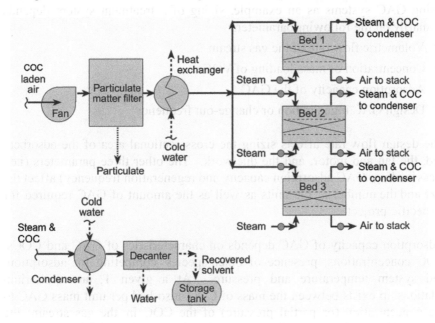

Figure 7.4 - A 3-bed fixed-bed regenerative adsorption system (Modified from EPA, 2012)

6.2.3 Factors affecting adsorption

In addition to the characteristics of the adsorbent (e.g., grain size, pore size, specific surface area, water vapor affinity, and polarity), a number of operating parameters will affect the performance of an adsorption system. Two pretreatment processes are often employed to the influent gas stream to enhance the treatment performance. The first is cooling, and the other is dehumidification. Adsorption of COCs is exothermic, which is favored by lower temperatures. In addition, at higher temperatures the vapor pressures of COCs would be higher so that the adsorbed COCs may have sufficient energies to overcome the attraction force between them and the adsorbent. As a rule of thumb, the waste gas stream needs to be cooled down to <130 °F (54 °C). On the other hand, water vapor molecules in the gas stream will compete with COCs for available adsorption sites. The RH of the influent gas stream should generally be reduced to <50%. For example, the off-gas stream from an air stripper is typically saturated with water. The gas stream needs to be cooled down (e.g., using chiller water) to condense out the moisture and then heated up to some extent (e.g., using an electrical heater) to raise its relative humidity before entering a GAC system.

6.2.4 Design considerations for adsorbers

Using GAC systems as an example, sizing of a treatment system depends primarily on the following parameters:

- Volumetric flow rate of the gas stream
- Concentration or mass loading of VOCs
- Adsorption capacity of the GAC
- Design GAC regeneration or change-out frequency

The design flow rate affects sizing the cross-sectional area of the adsorbent bed, the fan and motor, and the ductwork. The other three parameters (i.e., mass loading, GAC adsorption capacity and regeneration frequency) affect the size and the number of the units as well as the amount of GAC required for a specific project.

Adsorption capacity of GAC depends on characteristics of GAC and COCs, COC concentrations, presence of other species competing for adsorption, and system temperature and pressure. At a given T, an equilibrium relationship exists between the mass of COC adsorbed per unit mass GAC to the concentration (or partial pressure) of the COC in the gas stream; the relationship is often referred to as an *adsorption isotherm*. For most of the COCs, their adsorption isotherms can be fitted well by a power curve, also known as the *Freundlich isotherm*, as discussed in Chapter 3:

$$q = a(P_{COC})^m \qquad (7.1)$$

where q = equilibrium/saturation adsorption capacity (in kg COC/kg GAC, or lb COC/lb GAC), P_{COC} = partial pressure of VOC in the gas stream (in atm, or psi), and a, m = empirical constants. The empirical constants of Freundlich isotherms for selected VOCs can be obtained from vendors. It should be noted that values of these empirical constants are only for a specific type of GAC and for a specific COC and a concentration range at a given temperature.

The *saturation (equilibrium) capacity*, determined from the adsorption isotherm, is the maximum amount of the COC that the GAC can adsorb. The actual adsorption capacity in the field applications should be lower than the saturation capacity. Normally, design engineers take 25 to 50% of the equilibrium value as the *design/working capacity* (q_{design}) as a factor of safety.

The maximum amount of COCs that can be removed or held ($M_{removal}$) by a given amount of GAC (M_{GAC}) can be determined as:

$$M_{removal} = q_{design} \times M_{GAC} = q_{design} \times [V_{GAC} \times \rho_b] \qquad (7.2)$$

where V_{GAC} = volume of the GAC and ρ_b = bulk density of the GAC.

Example 7.1 Adsorption capacity of a GAC adsorber

Off-gas from a soil venting project is to be treated by GAC adsorbers. The m-xylene (C_8H_{10}) concentration in the off-gas is 800 ppm. The gas flow rate is 6 m³/min (210 ft³/min) at T = 20 °C. Two 500-kg (1,100-lb) GAC adsorbers are to be used. Determine the maximum amount of m-xylene that can be held by one GAC adsorber before exhausted. The adsorption isotherm (T = 20 °C) for m-xylene concentrations ranging from 0.0001 to 0.05 psi is (q in kg/kg and P in psi):

$$q = 0.527P^{0.0703}$$

Solution:

(a) Convert the concentration from ppm to psi as:

P = 800 ppm = 800 × 10⁻⁶ atm = 0.0118 psi

(d) Saturation capacity (q_{sat}) = $(0.527)(0.0118)^{0.0703}$ = 0.386 kg xylene/kg GAC

(e) Assuming design capacity is 40% of the saturation capacity, then

q_{design} = (40%)(q_{sat}) = (40%)(0.386) = 0.154 kg/kg

(f) Amount of xylene retained by the GAC before being exhausted

= (M_{GAC})(q_{design})

= (500 kg/unit)(0.154 kg xylene/kg GAC) = 77 kg xylene/unit.

Discussion:

1. The saturation capacity of vapor-phase GAC is typically less than 40% of its mass (i.e., < 0.4 kg COC/kg GAC) so that the typical design capacity is <0.2 kg/kg). For comparison, the adsorption capacity of liquid-phase GAC (used to water/wastewater treatment) is much smaller, typically in the order of 0.01 kg COC/kg GAC.

2. Care should be taken to ensure that the units of two concentrations in an adsorption isotherm go with the given empirical constants. In addition, the air pollutant concentration is within the applicable range.

3. The influent COC concentration in the gas stream, not the effluent concentration, should be used in an adsorption isotherm to determine the adsorption capacity.

To achieve efficient adsorption, the gas flow rate through the GAC adsorber should be kept low to allow for a sufficient contact time. The design gas velocity (v_{gas}) is often 18 m/min (60 ft/min), and 30 m/min (100 ft/min) is considered as the maximum. A minimal design velocity of 6 m/min (20 ft/min) is to avoid flow problems such as channeling (EPA, 2012).

The required cross-sectional area of the GAC adsorbers (A_{GAC}) can be readily determined as:

$$A_{GAC} = \frac{Q_{Gas}}{v_{gas}} \qquad (7.3)$$

where Q_{gas} is the volumetric influent gas flow rate.

For practical design considerations, there are minimum and maximum depths for the adsorbent bed. Figure 7.5 illustrates the mass transfer zone (MTZ), that is the zone of the bed where mass transfer occurs. When the leading edge of the MTZ reaches the end of the bed, the effluent COC concentration would start to rise rapidly, and is referred to as the *breakthrough point*. The design depth of the adsorbent bed is normally 0.6 m (2 feet) or deeper to accommodate the MTZ. The actual depth would also depend on the design change-out rate.

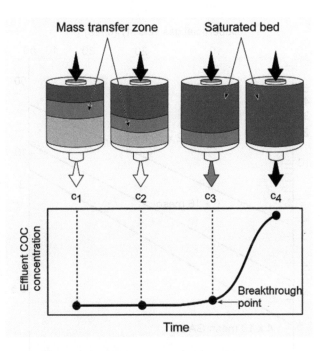

Figure 7.5 - Mass transfer zone (Modified from EPA, 2012)

The maximum practical bed depth is limited by the allowable pressure drop across the bed. Figure 7.6 shows that the pressure drop per unit depth increases with the superficial gas velocity and decreases with the size of GAC. With a gas velocity of 30 cm/s (18 m/min or 60 ft/min), the pressure drop will be about 5.7 kPa/m (7 in-H_2O/ft) using 6 × 16 mesh GAC (the middle line). If the total pressure drop across the bed is limited to < 6 kPa (24 in-H_2O), the bed length should be kept <1 m (3 ft).

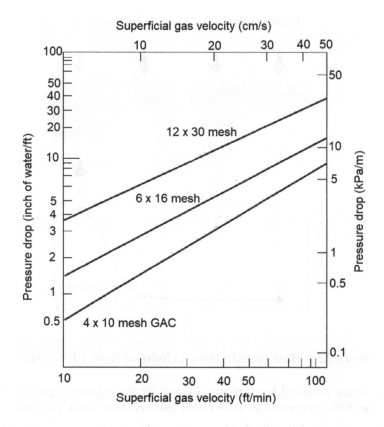

Figure 7.6 - Pressure drop versus superficial gas velocity through deep-bed GAC adsorbers (Modified from EPA, 2012)

Example 7.2 Required cross-sectional area of GAC adsorbers

Referring to the remediation project in Example 7.1, the 500-kg GAC units are out of stock. To avoid delay of remediation, off-the-shelf 55-gallon GAC units are proposed on an interim basis. The type of GAC in the 55-gallon (0.21 m³) units is the same as that in the 500-kg units. The vendor also provided the following information with regards to the units:

- diameter of GAC bed in each 55-gallon drum = 1.5 ft (0.5 m)

- depth of GAC bed in each 55-gallon drum = 3 ft (1.5 m)

- bulk density of the GAC = 28 lb/ft³ (449 kg/m³)

Determine (a) the mass of GAC in each 55-gallon unit, (b) the amount of xylene that each GAC unit can remove before being exhausted, and (c) the minimum number of the 55-gallon GAC units needed.

Solution:

(a) Volume of GAC inside a 55-gallon drum

$$= \pi(r^2)(h) = \pi[(1.5/2)^2](3) = 5.3 \text{ ft}^3 \ (= 0.15 \text{ m}^3)$$

Mass of the GAC inside a 55-gallon drum

$$= (V_{GAC})(\rho_b) = (5.3 \text{ ft}^3)(28 \text{ lb/ft}^3) = 148 \text{ lb} \ (= 67 \text{ kg})$$

(b) Mass of xylene that can be retained before the GAC becomes exhausted

$$= (M_{GAC})(q_{design}) = (148 \text{ lb/drum})(0.154 \text{ lb xylene/lb GAC})$$

$$= 22.8 \text{ lb xylene/drum} \ (= 10.4 \text{ kg xylene/drum})$$

(c) Assuming a design gas flow velocity of 60 ft/min (18 m/min), the required cross-sectional area of the GAC adsorbers can be found using Eq. 7.3 as:

$$A_{GAC} = \frac{Q_{Gas}}{v_{gas}} = \frac{210}{60} = 3.5 \ ft^2 \ (0.325 \ m^2)$$

If the adsorption system is tailor-made, then a system with a cross-sectional area of 3.5 ft² will do the job. However, the off-the-shelf 55-gallon drums are to be used. We need to determine the number of drums that will provide the required cross-sectional area.

Area of the GAC bed inside a 55-gallon drum

$$= \pi(r^2) = (\pi)[(1.5/2)^2] = 1.77 \text{ ft}^2/\text{drum} \ (= 0.164 \text{ m}^2/\text{drum})$$

Number of drums in-parallel to meet the design gas flow velocity

$$= (3.5 \text{ ft}^2) \div (1.77 \text{ ft}^2/\text{drum}) = 1.98 \text{ drums}$$

Using two drums in parallel would provide the required cross-sectional area to meet the design superficial gas velocity of 60 ft/min (18 m/min).

Discussion:

1. The bulk density of vapor-phase GAC is typically in the neighborhood of 30 lb/ft³ (480 kg/m³) or less. The amount of GAC in a 55-gallon (0.21 m³) drum is approximately 150 pounds (68 kg).

2. To meet the gas flow velocity requirement, the minimum number of 55-gallon drums needed for this project is two. Practically, the number of drums should be more to meet the monitoring requirements or the

desirable frequency of change-out. If multiple GAC adsorbers are used, the adsorbers are often arranged in series and in parallel. If two adsorbers are arranged in series, the monitoring point can be located at the effluent end of the first adsorber, while the effluent end of the second adsorber should serve as the sampling point for regulatory compliance. A high effluent concentration from the first adsorber indicates that it is reaching its capacity. The first adsorber should then be taken off-line, and the second adsorber shifted forward to be the first adsorber. Consequently, the capacity of both adsorbers can be fully utilized and the compliance can also be met. If there are two parallel trains of adsorbers, one train can always be taken off-line for regeneration or maintenance so that a continuous operation of the system can be ensured.

The COC removal rate by a GAC adsorber ($R_{removal}$) can be calculated by using the following formula:

$$R_{removal} = (G_{in} - G_{out})(Q_g) \qquad (7.4)$$

In practical applications, the effluent concentration (G_{out}) is kept below the discharge limit, which is often very low. Therefore, for a factor of safety, or as a conservative approach, the term of G_{out} can be deleted from Eq. 7.4 in design. The mass removal rate is then the same as the mass loading rate ($R_{loading}$):

$$R_{removal} \sim R_{loading} = (G_{in})(Q_g) \qquad (7.5)$$

The mass loading rate is the multiplication product of the gas flow rate and the mass concentration of the COC.

Example 7.3 Determine mass removal rate by GAC adsorbers

Referring to the remediation project described in Example 7.2, the discharge limit for xylene is 100 ppb. Determine the mass removal rate by the two 55- gallon GAC units.

Solution:

(a) MW of xylene (C_8H_{10}) = (12)(8) + (1)(10) = 106

At 20 °C (77 °F), 1 ppm = (106/385) $\times 10^{-6}$ lb/ft^3 = 2.75 $\times 10^{-7}$ lb/ft^3

= (106/24.05) mg/m^3 = 4.41 mg/m^3

800 ppm = (800)(2.75 $\times$ 10^{-7} lb/ft^3) = 2.2 $\times$ 10^{-4} lb/ft^3 (= 3.53 g/m^3)

(b) $R_{removal} \sim R_{loading} = (2.2 \times 10^{-4} \text{ lb/ft}^3)(210 \text{ ft}^3/\text{min}) = \underline{0.046 \text{ lb/min}}$

$$= (4.41 \text{ mg/m}^3)(6 \text{ m}^3/\text{min}) = \underline{0.021 \text{ kg/min}}$$

Once the GAC in an adsorber reaches its capacity, it should be regenerated or properly disposed of. The service time between two regenerations (or the service life of a fresh batch of GAC without regeneration), $t_{service}$, can be found by dividing the capacity of the GAC adsorber with $R_{removal}$ as:

$$t_{service} = \frac{M_{capacity}}{R_{removal}} = \frac{(M_{GAC})(q_{design})}{R_{removal}} \tag{7.6}$$

Example 7.4 Determine change-out frequency of GAC adsorbers

Referring to the remediation project described in Example 7.3, the discharge limit for xylene is 100 ppb. Determine the service life of the two 55-gallon GAC units.

Solution:

As shown in Example 7.2, the amount of xylene that each unit can retain before exhaustion is 22.8 lb (10.4 kg). The total service-life of these two units is:

$$t_{service} = \frac{M_{capacity}}{R_{removal}} = \frac{(2)(22.8 \text{ lb/unit})}{(0.046 \text{ lb/min})} = 991 \text{ min} < 1 day$$

Discussion:

1. Although two drums in parallel can provide a proper cross-sectional area, the relatively high COC concentration makes the service life of two 55-gallon GAC drums very short.

2. A 55-gallon GAC drum normally costs several hundred dollars. In this example, two drums last less than one day. With the labor and disposal costs, this alternative will be cost-prohibitive. A GAC system with on-site regeneration or other treatment alternatives should be considered.

If the COC concentration of the gas stream is high, a GAC adsorption system with on-site regeneration would become a more attractive option. The amount of GAC required for on-site regeneration depends on the mass loading rate, the GAC's adsorption capacity, the design service time between two regenerations, and the ratio between the number of GAC units in regeneration mode and the number of GAC units in adsorption mode. The total amount of GAC needed (M_{GAC}) can be determined by using the following formula (EPA, 1991):

$$M_{GAC} = \frac{(R_{removal})(t_{service})}{q_{design}}\left[1 + \frac{N_{des}}{N_{ads}}\right] \qquad (7.7)$$

where M_{GAC} = total amount of GAC required, $t_{service}$ = adsorption time between two consecutive regenerations, N_{ads} = number of GAC beds in the adsorption mode, and N_{des} = number of GAC beds in the regeneration (desorption) mode.

Example 7.5 Amount of GAC needed for on-site regeneration

Referring to the remediation project described in Example 7.3, an on-site re- generation GAC system is proposed to deal with the high COC mass loading. The system consists of three GAC adsorbers. Two of them will be in the adsorption mode while the other one in the regeneration mode. The adsorption cycle time is 24 hours. Determine the amount of GAC required for this system.

Solution:

The total amount of GAC required in all three adsorbers:

$$M_{GAC} = \frac{(0.046\ lb/min)(1440\ min)}{0.154\ lb/lb}\left[1 + \frac{1}{2}\right] = 645\ lb\ (293\ kg)$$

A total of 645 lb (293 kg) of GAC is needed (i.e., 215 lb (98 kg) in each bed).

7.3 Absorption

For air pollution control, absorption refers to removal of undesirable gaseous components from a gas stream by dissolving them into a liquid. It is a mass transfer process and the main removal mechanism is dissolution. The participants of this process include (a) *absorbent*: the liquid, usually water, into which the COC is absorbed/retained; (b) *absorbate:* the COC that is being absorbed into the absorbent; and (c) *carrier gas*: the inert part of the gas stream, typically air, from which the COCs are to be removed.

7.3.1 Principles of absorption

What is the main difference between adsorption and absorption? Adsorption is a surface phenomenon and the COCs are retained on the surface of the adsorbent, a solid. In absorption, the COCs in the gas phase go through the air-liquid interface, retained and distributed throughout the absorbent, a liquid. A two-film theory is commonly used to describe mass transfer of a compound across the interfacial boundary between the gas and the liquid. In this theory, there are a gas film on one side of the boundary and a liquid on the other side. They are stagnant films and hypothetical (Figure 7.7). The gas pollutant concentration in its bulk phase is G, while the pollutant concentration in the bulk liquid phase is C. The gas concentration (G) is greater than a concentration that would be in equilibrium with the liquid concentration (C). Therefore, the gaseous compound has a tendency to move into the liquid phase. Consequently, the gas concentration would decrease to G_i at the interface to be in equilibrium with C_i, the liquid concentration at the interface. The concentration gradients between the bulk phases and the interface are the driving forces for the mass transfer.

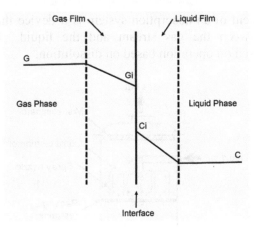

Figure 7.7 - Two-film theory for mass-transfer in absorption

The dissolution of the COCs into the liquid is the main removal mechanism. Solubility is the most important criterion to determine the applicability of the absorption process. Thus the adsorbers are applicable for relatively-soluble gaseous organic and many inorganic (e.g., H_2S, HCl, and NH_3) pollutants.

The solubility limit can be overcome by changing its pH and/or adding chemicals into the liquid to react with the dissolved gas pollutants. For example, raising the liquid pH would improve the effectiveness of an adsorber for removal of NH_3; adding an oxidant such as hypochlorite (OCl^-) for removal of reduced compounds; and adding lime, $Ca(OH)_2$, to react irreversibly with SO_2 to form $CaSO_4$ as in flue gas desulfurization (FGD).

7.3.2 Absorption systems and their components

Absorbers in air pollution control use a liquid to remove/scrub COCs from a gas stream and they operate under the principle of absorption. Since it is a mass-transfer process, a large air-liquid interfacial area and contact time available for mass transfer are the important factors affecting performances of absorbers or wet scrubbers. As mentioned, wet scrubbers/absorbers can simultaneously remove gaseous pollutants and particulates from a gas stream. However, the main removal mechanism for particulate removal and that for gaseous pollutants are different. The main mechanism for particulate removal is inertial impaction and the treatment systems are more often referred to as wet scrubbers. Dissolution is the main mechanism for removal of gaseous pollutants, and the removal systems are more often referred to as absorbers.

The main component of an absorption system is a device that enhances the close contact between the gas stream and the liquid. The subsequent discussion is focused on operation based on dissolution.

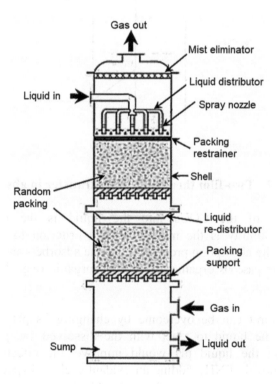

Figure 7.8 - Packed tower for gas absorption (Modified from EPA, 2002)

To provide enough contact time, an absorber is often a tall column in which the air flows counter-currently with the liquid. The common configurations include spray tower, packed column, and tray column. Packed column is the most common configuration for absorbers and it is the focus here. In addition to packing, the system consists of a liquid pump, an air blower, a liquid distributor (or sprayers) to distribute the liquid, and a mist eliminator to minimize the carry-over of liquid droplets existing the top of the column. There are also a packing support plate at the bottom and a restrainer at the top to keep the packing in place. A sump at the bottom of the column is also common for liquid storage and the liquid can be recirculated or added back as part of the scrubbing liquid (Figure 7.8).

Packing materials provide a large specific surface area, on which the liquid will form a thin layer, for mass transfer between the liquid and the gas to occur. A specific packing is described by its trade name and overall size (e.g., 2-inch Jaegar Tri-Packs®). Some common types of packing have been shown in Figure 6.32. Their overall dimensions normally range from 2.5 to 10 cm (1 to 4 inches). They are made of plastics (e.g., polyethylene (PE), polypropylene (PP)), ceramic, or metal. However, for air pollution control applications, plastic packing materials are most common. Main considerations in selection of the packing materials include size, temperature rating, corrosion resistance, bulk density, and cost. Figure 7.9 illustrates different sizes of Jaeger Tri-Packs®.

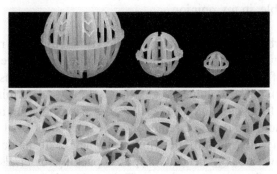

Figure 7.9 - Different sizes of Jaeger Tri-Packs® (http://www.jaeger.com/)

Cross-flow absorbers are also being used. For this type of configuration, the air stream flow flows horizontally through the packing materials, in which the scrubbing liquid is flowing down through the packing.

An air stripper operates on the same principle as an absorber, both based on the mass transfer of COCs between the liquid and air. Air strippers use clean air to treat dirty liquid, while absorbers use clean water to treat dirty air.

7.3.3 Factors affecting absorption

Henry's constant is the most important pollutant characteristics affecting the removal. A COC with a larger Henry's constant implies it prefers staying in the air. Therefore the absorption will work better for compounds with smaller Henry's constant. In addition, Henry's constants of most VOCs increase with temperature. Thus the extent of absorption of VOCs will be greater in lower temperatures.

Type and size of packing used will affect the mass transfer between the liquid and the gas stream. For a given type of packing, for example, a smaller packing size should provide a better mass transfer since it has a larger specific surface area. However, the trade-off is a higher pressure drop. A larger packing bed should also yield a better removal, but it comes with a higher capital and operating cost. The liquid-to-air ratio is another important operating parameter.

7.3.4 Design calculations

Absorption is also a very common process in chemical engineering practices. Design examples for absorbers in most, if not all, of the environmental engineering literature used the classical chemical engineering's approach in which the concentrations are expressed in mole fraction and graphical methods are used. Let us take a different approach, using the mass concentrations and no graphics. It is more straight forward and apprehensible, in my opinion.

Sizing an absorber depends primarily on the volumetric flow rate of the gas stream, the influent COC concentration, the required removal efficiency, and the solubility of the COC in the liquid (actually, Henry's constant will be a better indicator). The volumetric gas flow rate (Q_G) determines the cross-sectional area of the column (A) as well as the size of the fan and the motor. The concentration, the required removal, and Henry's constant determine the residence time (which, in turn, determines the height of the column) as well as the liquid flow rate (Q_L) needed. Types of packing materials which would have impacts on the mass transfer coefficient (in other words, the required residence time for a specific removal) and the head-loss across the column.

In a packed-column absorption tower, the polluted air and the liquid stream flow counter-currently through a packing column. The packing provides a large surface area for VOCs to migrate from the air stream to the liquid.

A mass balance equation can be derived by letting the amount of COCs removed from the gas stream be equal to the amount of the COCs retained by the liquid (Figure 7.10):

$$Q_G(G_{in} - G_{out}) = Q_L(C_{out} - C_{in}) \qquad (7.8)$$

where G = COC concentration in the air phase (mg/L), and C = COC concentration in the liquid phase (mg/L).

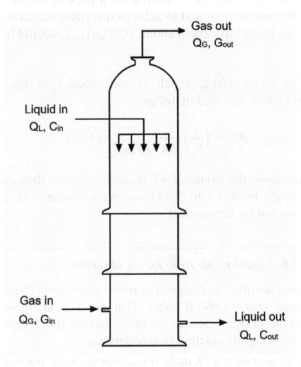

Figure 7.10 - Mass balance around an absorber
(Modified from EPA, 2003)

For an ideal case where the influent liquid contains no COCs (C_{in} = 0) and the effluent gas stream becomes completely free of COC (G_{out} = 0), Eq. 7.8 can be simplified to:

$$(Q_G)(G_{in}) = (Q_L)(C_{out}) \qquad (7.9)$$

Assuming that Henry's law applies and the influent gas stream is in equilibrium with the effluent liquid, then:

$$G_{in} = H^* \times C_{out} \qquad (7.10)$$

211

where H^* = Henrys constant of the COC in a dimensionless form. Combining the two equations above, the following relationship can be derived:

$$\left(\frac{1}{H^*}\right)\left[\frac{Q_L}{Q_W}\right]_{min} = 1 \qquad (7.11)$$

$(Q_L/Q_G)_{min}$ is the minimum liquid-to-air ratio (in vol/vol) which is the ratio for an ideal/theoretical case. It would require a packing height of relatively large to reach the equilibrium and to achieve the 100% removal. The actual liquid-to-air ratio in real-life applications, $(Q_L/Q_G)_{actual}$, should be larger than the $(Q_L/Q_G)_{min}$.

The *absorption factor* (AF), which is the product of the $(1/H^*)$ and $(Q_L/Q_G)_{actual}$, is commonly-used in design:

$$AF = \left(\frac{1}{H^*}\right)\left[\frac{Q_L}{Q_W}\right]_{actual} \qquad (7.12)$$

For field applications, the values of AF should be greater than one. Practical values of AF range from 1.5 to 2. Operating a system with an AF value larger than 2 may not be economical.

Example 7.6 Liquid-to-air ratio for an absorber

A packed-column absorber is designed to remove ammonia (NH_3) from a gas stream ($Q = 85$ m³/min = 3,000 ft³/min). Determine (a) the minimum liquid-to-air ratio, (b) the design liquid-to-air ratio, and (c) the design liquid flow rate. Use the following information in calculations:

- Henry's constant = 0.78 mole fraction in air/mole fraction in water
- absorption factor (AF) = 1.5
- T = 298 K and P = 1 atm.

Solution:

(a) Convert the given Henry's constant into the dimensionless value:

H^* (dimensionless)

= H [in (mole frac. in air)/(mole frac. in liq.)]×W÷ [R×T×γ×1,000]×P

= [(0.78)(18)] ÷ [(0.082)(298)(1)(1000)] × 1 = 5.75×10⁻⁴

$$\left(\frac{1}{H^*}\right)\left[\frac{Q_L}{Q_W}\right]_{min} = 1 = \left(\frac{1}{5.75 \times 10^{-4}}\right)\left[\frac{Q_L}{Q_W}\right]_{min}$$

So $(Q_L/Q_G)_{min} = 5.75 \times 10^{-4}$

(b) $(Q_L/Q_G)_{actual} = (Q_L/Q_G)_{min} \times AF = (5.75 \times 10^{-4})(1.5) = 8.62 \times 10^{-4}$

(c) $Q_L = (85)(8.62 \times 10^{-4}) = 7.33 \times 10^{-2}$ m^3/min = 19.4 gal/min

Discussion: An absorption factor value of 1.5 means the $(Q_L/Q_G)_{actual}$ ratio is 1.5 times of the $(Q_L/Q_G)_{min}$ ratio. This, the $(Q_L/Q_G)_{actual}$ can be obtained by multiplying the $(Q_L/Q_G)_{min}$ with this AF value.

One of the key components in sizing an absorber is to determine the diameter of the column. The diameter depends mainly on the gas flow rate. The larger the volumetric gas flow rate, the larger the column diameter would be. For a given cross-sectional area and a specific liquid flow rate, if the gas flow velocity increases to a level, the liquid may start to be held up in the void of the packing bed. Due to reduction in the available channels for gas travel through the bed, the pressure drop across the bed will increase and the mass transfer between the gas and the liquid adversely decreases.

A further increase in the gas velocity, the liquid will stop flowing altogether and collects on the top of the packing bed. This condition is called "flooding," and the relevant gas velocity is called the flooding velocity. The flooding velocity depends mainly on the density of the gas stream, density and viscosity of the liquid, and type of the packing. It is advisable to get specific monographs from the vendor of the packing materials to derive the site-specific flooding velocity. The design gas velocity (v_G) is chosen to be 50 to 75% of the flooding velocity. This velocity is then used to determine the required cross-sectional area of the stripping column ($A_{absorber}$):

$$A_{absorber} = \frac{Q_G}{v_G} \qquad (7.13)$$

The required depth of the packing column (Z) for a specific removal efficiency is another important design component. A taller column would be required to achieve a larger removal efficiency. The packing height can be determined using the classical chemical engineering's transfer unit concept:

$$Z = HTU \times NTU \qquad (7.14)$$

where HTU = height of transfer unit and NTU = number of transfer units.

The HTU value depends on the gas flow velocity and the overall gas-phase mass transfer coefficient, K_Ga in which K_G is the mass transfer rate constant (m/sec) and a is the specific surface area (m^2/m^3). K_Ga has a unit of 1/time. The K_Ga value for a specific application should be best determined from pilot testing. There are also empirical equations available to estimate K_Ga values. HTU has a unit of length and can be estimated by:

$$HTU = \frac{v_G}{K_Ga} \qquad (7.15)$$

The NTU value can be determined by using the following formula:

$$NTU = \left(\frac{AF}{AF-1}\right) \ln\left\{\left(\frac{G_{in}-H^* \times C_{in}}{G_{out}-H^* \times C_{in}}\right)\left(\frac{AF-1}{AF}\right) + \frac{1}{AF}\right\} \qquad (7.16)$$

The equation can be simplified if the influent liquid is free of COC, by having $C_{in} = 0$, as

$$NTU = \left(\frac{AF}{AF-1}\right) \ln\left\{\left(\frac{G_{in}}{G_{out}}\right)\left(\frac{AF-1}{AF}\right) + \frac{1}{AF}\right\} \qquad (7.17)$$

The equation can be further simplified for special cases where a chemical reaction occurs once the COC enters the liquid or if the COC is extremely soluble. For these two cases, the COC will exhibit insignificant vapor pressure so that the Henry's constant is essentially zero. With that,

$$NTU = \ln\left(\frac{G_{in}}{G_{out}}\right) \qquad (7.18)$$

For this case, NTU for a 90% reduction in the influent COC concentration would be equal to $\ln 10$ (= 2.303).

Example 7.7 Sizing an absorber

For the absorber in Example 7.6, if the influent ammonia concentration is 3% and the required removal is 90%. The diameter of the column is 0.75 m and $K_Ga = 6.5$/s. Determine the packing height required.

Solution:

(a) Cross-sectional area = $\pi(0.75/2)^2 = 0.442$ m^2

(b) Use Eq. 7.13,

$$v_g = \frac{85}{0.442} = 192.4 \; m/min = 3.21 \; m/s$$

214

(c) Use Eq. 7.15,

$$HTU = \frac{3.21 \, m/s}{6.5/s} = 0.49 \, m$$

(d) Use Eq. 7.17,

$$NTU = \left(\frac{1.5}{1.5 - 1}\right) ln\left\{\left(\frac{0.03}{0.003}\right)\left(\frac{1.5 - 1}{1.5}\right) + \frac{1}{1.5}\right\} = 4.16$$

(e) Use Eq. 7.14,

$$Z = HTU \times NTU = (0.49 \, m)(4.16) = 2.05 \, m$$

7.4 Thermal Oxidation

Combustion is the process of burning. For a more scientific definition, combustion is a chemical reaction process in which a substance reacts rapidly with oxygen at an elevated temperature. The substance is in a more reduced form and often called the *fuel*, while the source of oxygen is called the *oxidizer*. The fuel is typically an organic compound that possesses a *heating value*. The combustion is a destruction process in which the fuel gets destroyed and the process is exothermic (i.e., energy is released from the reaction). It is often called a *thermal oxidation* process. We burn fossil fuels, such as coal, natural gas, and gasoline to generate energy for many daily activities. *Incineration* is a waste treatment process that destroys organic substances in a waste solid, liquid, or gas stream.

With regards to air pollution control, thermal oxidation processes are commonly used to treat gas streams containing VOCs and/or toxic organic compounds. Thermal oxidation is also applicable to some inorganic compounds such as H_2S. They are often grouped into (direct) thermal oxidation and catalytic oxidation. Both groups are operated under the same principle which employs elevated temperatures to oxidize/destroy organic compounds, except catalysts are used in catalytic incineration. The section will focuses on (direct) thermal oxidation, while the next section is on catalytic oxidation.

7.4.1 Principle of thermal oxidation

Most fuels used for combustion are composed essentially of carbon and hydrogen. A simplified reaction for thermal oxidation of hydrocarbons can be expressed as:

$$C_xH_y + \left(x + \frac{y}{4}\right)O_2 \rightarrow x \, CO_2 + \left(\frac{y}{2}\right)H_2O + Heat \qquad (7.19)$$

As shown, the reaction products from combusting hydrocarbons are carbon dioxide and water. For methane, as an example,

$$CH_4 + 2O_2 \rightarrow CO_2 + 2H_2O \quad (7.20)$$

It requires two moles of oxygen to completely combust one mole of methane. The quantity of oxygen needed for a complete oxidation is called the *stoichiometric amount* or *theoretical amount*. The stoichiometric ratio of O_2:CH_4 is 2:1. Since air, instead of pure oxygen, is used in combustion, the stoichiometric ratio of air:CH_4 would be approximately 9.52:1, assuming the air contains 21 oxygen (i.e., $9.52 = 2 \div 21\%$). If the amount of oxygen is less than the stoichiometric amount to burn all the fuel, the air-fuel mixture is referred to as *rich*. The combustion is often called *rich burning*. When the oxygen amount is more than the stoichiometric amount, the air-fuel mixture will be *lean* and combustion is called *lean burning*.

The minimum concentration of a combustible vapor to support its combustion is defined as its *Lower Explosive Limit* (*LEL*). Below this level, the mixture is "too lean" to burn. The maximum concentration of a compound that will burn in the air is defined as its *Upper Explosive Limit* (*UEL*). Above this level, the mixture is "too rich" to burn. The range between the LEL and UEL is the *flammable range* for that compound. Table 7.2 tabulates the LEL, UEL, and *stoichiometric ratios* of some common compounds.

Table 7.2 - LEL, UEL, and Stoichiometric Mixture (% by volume)

Compound	LEL	UEL	Stoichiometric Mixture
Acetone	2.5	12.8	4.97
Benzene	1.2	7.8	2.71
Ethane	2.8	15.3	5.64
Hydrogen	2.0	80.0	29.5
Methane	4.4	15.5	9.48
Octane	1.0	6.5	1.65
Pentane	1.4	9.2	2.55
Propane	2.1	9.5	4.02
Toluene	1.1	7.1	2.27
Xylenes	0.9	7.0	1.96
Gasoline	1.4	7.6	varies w/ compositions

The LEL and UEL values in Table 7.2 and those found in literature should be used with caution [Note: the reported values are often different]. It is because they are measured values, the data might be derived at different T, P, and oxygen concentrations . Because of uncertainty in the LEL values and in the accuracy of real-time monitoring, the influent concentrations to the gas control system are usually kept to be less than 25% of the LEL to provide some margin of protection from fires and explosion within the system. The LEL and UEL values of a gas mixture, without having actual laboratory measurements, can be estimated as the weighted average of each combustible constituents in the gas stream as:

$$LEL_{mixture} = \frac{1}{[\Sigma(y_i/LEL_i)]} \quad \& \quad UEL_{mixture} = \frac{1}{[\Sigma(y_i/UEL_i)]} \tag{7.21}$$

where y_i = fraction of combustible component i of all the combustible components. For a more conservative approach one can use the lowest LEL of these combustible compounds as the LEL of the mixture (EPA, 2012).

Most gas compositions in combustion calculations are expressed in volume %; however, weight % is sometime used such as the "air/fuel" ratio (A/F) for internal combustion (IC) engines. The A/F ratio for an air-fuel mixture can be found as:

$$A/F = \frac{n_{air}MW_{air}}{n_{fuel}MW_{fuel}} \tag{7.22}$$

where n_{air} is the number of moles of air needed to stoichiometrically combust a specific number of moles of the fuel (n_{fuel}).

Example 7.8: LEL of a gas mixture

A gas stream contains 1,250 ppm of benzene, 750 ppm of toluene and 500 ppm of xylenes. Does it exceed 25% of the LEL for the entire gas stream?

Solution:

(a) From Table 4.2, the LELs of benzene, toluene, and xylenes are 1.2, 1.1, and 0.9%, respectively.

(b) With a total concentration of 2,500 ppm (1,250 + 750 + 500), the fractions of benzene, toluene and xylenes are 0.5, 0.3 and 0.2, respectively.

$$LEL_{mixture} = \frac{1}{\left[\dfrac{0.5}{1.2\%} + \dfrac{0.3}{1.1\%} + \dfrac{0.2}{0.9\%}\right]} = 1.10\% = 11,000\ ppmV$$

25% of the $LEL_{mixture}$ = (25%)(11,000) = 2,750 ppm > 2,500 ppm

The total concentration is less than the 25% LEL value.

(c) For a more conservative approach:

The lowest LEL of these three compounds is 0.9% (xylenes),

25% of 9,000 ppm = (25%)(9,000) = 2,250 ppm < 2,500 ppm

The total concentration exceeds the 25% LEL value.

Discussion: 1% (by volume) = (1/100) = (10,000/1,000,000) = 10,000 ppm

---

Example 7.9: Stoichiometric mixture and A/F ratio

Determine methane concentration (in %) in a stoichiometric mixture for combustion and the corresponding A/F ratio.

Solution:

(a) As shown in the text, the stoichiometric ratio of air to methane is 9.52.

$$Concentration\ of\ CH_4 = \frac{n_{methane}}{n_{methane} + n_{air}} = \frac{1}{1 + 9.52} = 9.5\%$$

(b) $\dfrac{A}{F} = \dfrac{n_{air}MW_{air}}{n_{fuel}MW_{fuel}} = \dfrac{(1-9.5\%)(29)}{(9.5\%)(16)} = 17.3\ \dfrac{kg\ air}{kg\ methane} = 17.3\ \dfrac{lb\ air}{lb\ methane}$

Discussion: The LEL, UEL, and stoichiometric mixture concentrations (in %) vary considerably with molecular weight of the fuel; however, they vary much less on a weight basis (e.g., the A/F ratio).

7.4.2 Thermal oxidation systems and their components

A thermal oxidizer has a chamber that has gas- or oil-fired burners. The burners are used to heat up the gas stream to exceed the auto-ignition temperatures of the COCs. *Auto-ignition temperature* is the temperature at which a compound spontaneously ignites without an external source of ignition in a normal atmosphere (while *flash point* is the lowest temperature at which vapor of a volatile compound can be ignited in the presence of an ignition source). Auxiliary fuel and auxiliary air are often needed in thermal oxidation. The thermal oxidation reaction will then occur in the combustion chamber which is lined with refractory materials for insulation. Although a heat recovery system is not mandatory, a heat recovery device is commonly employed to recover some heat/energy carried by the flue gas before it is discharged into the atmosphere (Figure 7.11).

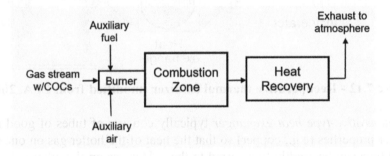

Figure 7.11 - Schematic of thermal oxidation system

Heat of hot flue gas can be recovered by preheating the influent stream to the thermal oxidizer (i.e., *primary heat recovery*) or heating up other processes at the site (i.e., *secondary heat recovery*). Although direct recycling the flue gas into the combustion chamber is theoretical feasible, it would increase the total flow rate to the combustion chamber to reduce the residence time.

Recuperative and regenerative devices refer to the type of heat exchangers used to enhance the heat recovery efficiency. Figure 7.12 illustrates a thermal oxidation system equipped with a double-pass recuperative heat exchanger. In this system, the COC-laden gas (T = 120 °C = 248 °F) is heated up to 351 °C (664 °F) while the flue gas from the combustion zone (T = 700 °C = 1,292 °F) is cooled down to 315 °C (599 °F) before moving to the stack for discharge.

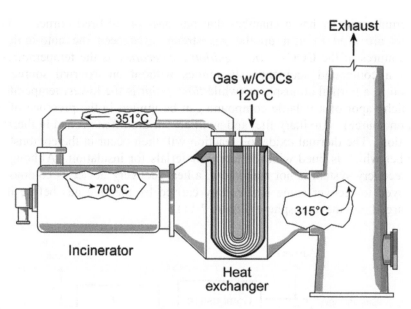

Figure 7.12 - Recuperative thermal oxidizer (modified from EPA, 2003)

A *recuperative-type heat exchanger* typically consists of tubes of good heat transfer properties (e.g., copper) so that the heat of the hotter gas on one side of the tube can be readily transferred to the colder gas on the other side (see Figure 7.13 for a shell-tube heat exchanger). The *regenerative heat recovery* devices use heat absorbing materials to store heat from a hot gas stream and the heat will be released to a cooler stream later. It should be noted that there are at least two limitations for the maximum heat recovery. Firstly, the temperature of the hot gas should not be cooled below the dew point of any gases because the condensate may be corrosive. Secondly, the temperature of the cooler stream should not be raised above its auto-ignition temperature.

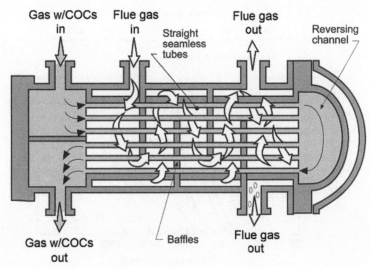

Gas w/COCs in   Flue gas in   Straight seamless tubes   Flue gas out   Reversing channel

Gas w/COCs out   Baffles   Flue gas out

Figure 7.13 - Schematic of a recuperative shell-tube heat exchanger (modified from EPA, 2003)

7.4.3 Factors affecting thermal oxidation

In addition to sufficient oxygen concentrations, an efficient combustion process requires three T's: a sufficiently high temperature, an adequate residence time, and a good contact between the fuel and the oxidizer (good turbulence) in the combustion chamber. These three parameters govern the speed and completeness of thermal oxidation. It should be noted that they are independent variables. The rate of COC oxidation is greatly influenced by the temperature in the combustion chamber. The higher the temperature, the faster the oxidation reaction proceeds. Figure 7.14 illustrates the effects of combustion temperature and residence time on destruction of a hypothetical COC. For example, the same destruction efficiency can be achieved at a higher temperature with a shorter residence time, or at a lower temperature with a longer residence time. In addition, a higher temperature would yield a high destruction efficiency.

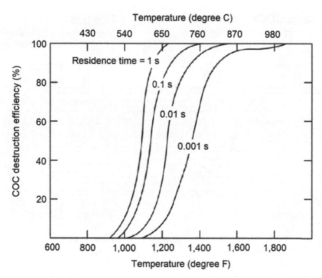

Figure 7.14 - Effects of temperature and residence time on COC destruction (modified from EPA, 2012)

Table 7.3 tabulates the auto-ignition temperatures of several organic compounds. As shown, they range from 869 to 1,245 °F (465 to 674 °C). The combustion temperatures for efficient oxidation is generally 200 to 500 °F (~110 to 280 °C) above the auto-ignition temperature of the COC that is most difficult to be oxidized in the gas stream. The minimum thermal oxidation temperature is ~1,300 °F (~700 °C); and most of the organic vapors can be oxidized at this temperature. However, the equilibrium conditions among CO, O_2, and CO_2 would increasingly favor CO, instead of CO_2, as the combustion product, if the temperature is below 1,300 °F (~700 °C). Consequently, most of the thermal oxidizers operate at temperatures ranging from 1,300 to 1,800 °F (~700 to 1,000 °C) (EPA, 2003).

Table 7.3 - Auto-ignition temperatures (Modified from EPA, 2003)

Compound	Temperature	
	(°F)	(°C)
Acetone	869	465
Benzene	1,097	592
Chlorobenzene	1,245	674
Methanol	878	470
Methyl ethyl ketone (MEK)	960	516
Toluene	997	536
Xylene	924	496

If the waste air stream has a low oxygen content (below 13 to 16 percent), then auxiliary air would be needed to raise the oxygen level. It is to ensure flame stability of the burner and to ensure the concentration of CO in the flue gas is low.

The residence time needed to achieve a specific destruction efficiency in the combustion chamber depends partly on the rate of the COC oxidation at the prevailing temperature and partly on the degree of mixing between the gas stream and the auxiliary air in the combustion zone. To promote mixing, baffles can be installed within the combustion chamber to increase the turbulence. Residence times for typical thermal oxidizers are between 0.3 to 2 seconds (EPA, 2002).

For incineration of hazardous waste containing principal organic hazardous constituent (POHC), *destruction and removal efficiency* (DRE) is used as an indicator of removal efficiency as:

$$DRE\ (\%) = \frac{W_{in} - W_{out}}{W_{in}} \times 100 \qquad (7.23)$$

where W_{in} = influent mass flow rate of a particular POHC and W_{out} = its effluent rate in mass/time.

In addition, combustion efficiency (CE) is calculated by using volume concentrations of CO and CO_2 in the flue gas as:

$$CE\ (\%) = \frac{[CO_2]}{[CO_2] + [CO]} \times 100 \qquad (7.24)$$

Higher DRE and CE values are usually more desirable.

7.4.4 Design calculations for thermal oxidizers
The design of a thermal oxidation system depends on the COC concentrations in the influent gas stream, types of COCs, and the required destruction efficiencies or the required effluent COC concentrations. It is generally easier to achieve a specific destruction efficiency if the influent COC concentrations are higher.

Higher COC concentrations in the influent gas stream would help to raise the combustion temperature because of the larger heat value. That is one of the reasons why thermal incineration is a favorable alternative for treating a gas stream with high organic concentrations. However, for gas streams containing hazardous air pollutants, influent concentrations of flammable

223

vapors to a thermal oxidizer are generally limited to 25% of the LEL, imposed by insurance companies for safety concerns. Vapor concentrations up to 40 to 50% of the LEL may be permissible, if on-line monitoring of VOC concentrations and automatic process control and shutdown are employed. Flares are used for combustion of gas streams having vapor concentrations >100% of the UEL (EPA, 2012).

The volumetric gas flow rate is commonly expressed in ft^3/min in the US customary system that is cubic feet per minute (cfm). Since the volumetric flow rate of an air stream is a function of temperature and the gas stream undergoes zones of different temperatures in a thermal process, the air flow rates are shown in actual cfm (acfm) or standard cfm (scfm). The acfm refers to the volumetric flow rate under the actual temperature and pressure, while scfm is the flow rate at the standard conditions. The standard conditions are the basis for comparison. Unfortunately, as mentioned earlier the definition of the standard conditions is not universal. For EPA's documents related to combustion, the standard conditions are T = 77 °F (25 °C) and P =1 atm. Consequently, this standard temperature is used for combustion calculations here.

Conversions between an actual flow rate and its corresponding standard flow rate can be readily done by using the formula below. It assumes that the ideal gas law is valid and the pressure stays relatively constant:

$$\frac{Q_{actual}}{Q_{standard}} = \frac{273 + T_{actual}(in\ {}^oC)}{273 + 25} = \frac{460 + T_{actual}(in\ {}^oF)}{460 + 77} \quad (7.25)$$

Example 7.9: Conversion between actual and standard flow rates

A thermal oxidizer is used to treat off-gas from an industrial process. To achieve the required removal efficiency, the oxidizer operates at 700 °C (1,292 °F). A heat exchanger is employed to recuperate heat (see Figure 7.12) and the temperature to the stack is 315 °C (599 °F). The actual flow rate at the exit of the thermal oxidizer is 1,000 actual m^3/min (35,300 acfm).

(a) What would be the flow rate at the exit of the thermal oxidizer under standard conditions?

(b) Determine the exit velocity from the stack (diameter = 1 m = 3.28 ft).

Solution:

(a) Use Eq. 7.25 to convert the actual flow rate at the exit of the thermal oxidizer to its standard flow rate:

$$\frac{1{,}000}{Q_{standard}} = \frac{273+700}{273+25} = \frac{460+1{,}292}{460+77}$$

$Q_{standard} = 385.5$ standard m^3/min (13,610 acfm)

(b) Use Eq. 7.25 again to convert the standard flow rate to the actual flow rate at the stack:

$$\frac{Q_{actual}}{385.3} = \frac{273+315}{273+25} = \frac{460+599}{460+77}$$

$Q_{actual} = 761$ standard m^3/min (26,840 acfm)

Exit velocity = Q/A = (761 m^3/min) ÷ $[\pi(1)^2/4]$ = 970 m/min = 3,200 ft/min

Discussion: If the actual flow rate at one temperature is known, the actual flow rate at another temperature can be directly determined by using:

$$\frac{Q_{actual@T_1}}{Q_{actual@T_2}} = \frac{273 + T_1(in\ ^oC)}{273 + T_2(in\ ^oC)} = \frac{460 + T_1(in\ ^oR)}{460 + T_2(in\ ^oR)} \qquad (7.26)$$

The reported heating value of a COC is often on the basis of mass. It the data are not available, Dulong's formula can be used (Eqs. 2.19 & 2.20). Heat content of a gas stream is the sum of heat contents of all the COCs contained:

$$Heat\ content\ of\ a\ gas\ tream\ \left(\frac{kJ}{m^3}\ or\ \frac{Btu}{scf}\right) =$$
$$\sum\left\{Heat\ content\ of\ COC\ \left(\frac{kJ}{kg}\ or\ \frac{Btu}{lb}\right) \times COC\ conc.\left(\frac{kg}{m^3}\ or\ \frac{lb}{scf}\right)\right\} \quad (7.27)$$

Heat content of a gas stream in Btu/lb or kJ/kg can be converted from the corresponding value in Btu/ft^3 or kJ/m^3 by using its density:

$$Heat\ content\ of\ a\ gas\ stream\ \left(\frac{kJ}{kg}\ or\ \frac{Btu}{lb}\right) =$$
$$Its\ heat\ content\ \left(\frac{kJ}{m^3}\ or\ \frac{Btu}{ft^3}\right) ÷ its\ density\ \left(\frac{kg}{m^3}\ or\ \frac{lb}{ft^3}\right) \qquad (7.28)$$

Example 7.10: Estimate heating value of a gas stream

Referring to the project described in Example 7.3, a thermal oxidizer is also considered an alternative to treat the off-gas. Estimate the heat content of the gas stream that contains 800 ppm of xylene.

Solution:

(a) Use Dulong's formula (Eq. 2.19) to estimate the heating value of xylene:

$$\text{MW of xylene } (C_8H_{10}) = 12 \times 8 + 1 \times 10 = 106$$

$$\text{Weight percentage of } C = (12 \times 8) \div 106 = 90.57\%$$

$$\text{Weight percentage of } H = (1 \times 10) \div 106 = 9.43\%$$

$$\text{Heat Content } \left(\text{in } \frac{\text{Btu}}{\text{lb}}\right) = (145.4)(90.57) + (620)(9.43) = 19,015$$

(b) To determine the heating value of the gas stream containing 800 ppm xylene, we need to determine the mass concentration of xylene, which has been previously determined in Example 7.3:

$$800 \text{ ppm of xylene} = 2.2 \times 10^{-4} \text{ lb /ft}^3 = 3.53 \text{ g/m}^3$$

$$\text{Heat content} = (19,015 \text{ Btu/lb})(2.2 \times 10^{-4} \text{ lb/ft}^3) = 4.11 \text{ Btu/ft}^3$$

(c) Use Eq. 7.28 to convert the heat content from Btu/ft^3 to Btu/lb:

$$\text{Heat content} = (4.11 \text{ Btu/ft}^3) \div (0.0739 \text{ lb/ft}^3) = 55.6 \text{ Btu/lb}$$

Discussion:

1. Heat content of xylenes calculated from the Dulong's formula, 19,015 Btu/lb, is similar to that found in literature, 18,650 Btu/lb.

2. Using Dulong's formula in SI, one can easily find the corresponding values of 44,190 kJ/kg, 153 kJ/m^3, and 129 kJ/kg for (a), (b), and (c).

3. The density of the gas stream was assumed to be the same as that of the air. At 77 °F (25 °C), density of air = (29 lb)/(392 ft^3) or (29 g)/(24.46 L).

When a gas stream has a vapor concentration larger than 25% of the LEL, dilution air must be used to lower the COC concentrations to <25% of its LEL prior to combustion. The 25% LEL corresponds to a heat content of 176 Btu/lb or 13 Btu/scf (409 kJ/kg or 484 kJ/m^3) in most cases (EPA, 1991).

Example 7.11: Heat content of a gas stream at 25% of its LEL

A gas stream contains an elevated level of toluene (C_7H_8). The heat content of toluene is 18,500 Btu/lb (43,000 kJ/kg). Estimate the heat content of this gas stream corresponds to 25% of its LEL.

Solution:

(a) From Table 7.2, the LEL of toluene in air is 1.1% by volume.

Then, 25% of the LEL = (25%)(1.1%) = 0.275% by volume = 2,750 ppm

MW of toluene (C_7H_8) = $12 \times 7 + 1 \times 8 = 92$

1 ppm = $(92/392) \times 10^{-6} = 0.235 \times 10^{-6}$ lb/ft^3

$= (92/24.46) = 3.755$ mg/m^3 @ T = 77 °F (25 °C)

$2,750$ ppm = $(2,750)(0.235 \times 10^{-6}) = 6.46 \times 10^{-4}$ lb/ft^3 $(= 10.33$ g/m$^3)$

(b) Heat content = $(18,500$ Btu/lb$)(6.46 \times 10^{-4}$ lb/ft$^3) = \underline{11.95 \text{ Btu/ft}^3}$

$= (43,000$ kJ/kg$)(10.33 \times 10^{-3}$ kg/m$^3) = \underline{444.2 \text{ kJ/m}^3}$

(c) Heat content = $(11.95$ Btu/ft$^3) \div (0.0739$ lb/ft$^3) = \underline{161.7 \text{ Btu/lb}}$

$= (444.2$ kJ/m$^3) \div (1.184$ kg/m$^3) = \underline{375.3 \text{ kJ/kg}}$

Discussion: The calculated heating values are in line with those of typical VOCs at their 25% LELs.

When dilution is required, the volumetric flow rate of the dilution air can be found as (EPA, 1991):

$$Q_{dilution} = \left[\left({}^{H_w}/_{H_i} \right) - 1 \right] Q_w \qquad (7.29)$$

where

$Q_{dilution}$ = required flow rate of dilution air

Q_w = flow rate of the waste gas stream

H_w = heat content of the waste gas stream

H_i = heat content of the desired influent entering the combustion chamber

Example 7.12: Determine the amount of dilution air

The waste gas stream described in Example 7.11 is to be treated by direct thermal oxidation. The heat content of the gas stream was determined to be 700 kJ/kg (300 Btu/lb). The insurance policy limits the COC concentration in the influent to the combustion chamber to be ≤25% of its LEL. The flow rate of the gas stream is 20 standard m^3/min (706 scfm). Estimate the required dilution air flow rate.

Solution:

Use 375 kJ/kg (162 Btu/lb), from Example 7.11, as the heat content that corresponds to 25% LEL. Use Eq. 7.29:

$$Q_{dilution} = \left[\left(700/375\right) - 1\right](20) = 17.3 \, m^3/min \, (= 610 \, scfm)$$

If the waste air stream has a low oxygen content (below 13 to 16 percent), then supplemental air is needed to raise the oxygen level to ensure flame stability. If the exact composition of the waste gas stream is known, one can determine the stoichiometric amount of air (oxygen) for complete combustion. In general practices, excess air is added to ensure complete combustion. The following example illustrates how the stoichiometric amount of air and excess air for combusting a landfill gas are determined.

Example 7.13: Stoichiometric and supplemental air for combusting landfill gas

A landfill gas stream (Q = 100 standard m^3/min = 3,530 scfm) is to be treated by a thermal oxidizer. The landfill gas is composed of 60% by volume CH_4 and 40% CO_2. The gas is to be burned with 20% excess air at 950 °C (1,742 °F). Determine

(a) the stoichiometric amount of air required

(b) the supplemental air required

(c) the total influent flow rate to the thermal oxidizer

(4) the total effluent flow rate from the thermal oxidizer.

Solution:

(a) Influent Q of $CH_4 = (60\%)(100) = 60$ standard m^3/min

Influent Q of $CO_2 = (40\%)(100) = 40$ standard m^3/min

The reaction for complete combustion of methane is:

$$CH_4 + 2\,O_2 \rightarrow CO_2 + 2\,H_2O$$

The stoichiometric amount of $O_2 = (60)(2/1) = 120$ standard m^3/min

The stoichiometric amount of air $= (120)/(21\%) = 570$ standard m^3/min

(b) Q of O_2 in the influent $= (1 + 20\%)(120) = 144$ standard m^3/min

Q of the supplemental air $= (1 + 20\%)(570) = 684$ standard m^3/min

Q of N_2 in the supplemental air $= (79\%)(684) = 540$ standard m^3/min

(c) Total influent Q $= 60\ (CH_4) + 40\ (CO_2) + 684$ (supplemental air)

$\quad = 784$ standard m^3/min

(d) Q of O_2 in the effluent $= (20\%)(120) = 24$ standard m^3/min (20% excess)

Q of N_2 in the effluent $=$ Q of N_2 in the influent $= 540$ standard m^3/min

Q of CO_2 in the effluent $= CO_2$ in the landfill gas $+ CO_2$ from

$\quad$ combustion $= 40 + 60\ (CH_4:CO_2 = 1:1) = 100$ standard m^3/min

Q of water vapor in the effluent $=$ water vapor produced from

$\quad$ combustion $(CH_4:H_2O = 1:2) = (2)(60) = 120$ standard m^3/min

Total effluent Q $= 24 + 540 + 100 + 120 = 784$ standard m^3/min

Discussion:

1. The following table summarizes the flow rate of each component in this process:

	CH_4	O_2	N_2	CO_2	H_2O
Influent	60	$2(60)(1.2) = 144$	540	40	0
Effluent	0	$144 - 120 = 24$	540	$40 + 60$	120

2. Flow rates of the influent and the effluent are the same at 784 m^3/min under the standard conditions.

If COC concentrations in the waste gas stream are low, auxiliary fuel is needed to raise the combustion temperature to a desirable range. The following equation can be used to determine the amount of supplementary fuel required (EPA, 1991):

$$Q_{sf} = \frac{\rho_{wg}Q_{wg}[C_p(1.1T_{com}-T_{he}-0.1T_{ref})-H_{wg}]}{\rho_{sf}[H_{sf}-1.1C_p(T_{comb}-T_{ref})]}$$ [Eq. 7.30]

where

Q_{sf} = flow rate of the supplementary fuel (scfm or standard m^3/min)

ρ_{wg} = density of the waste gas stream (usually 0.0739 lb/ft^3 or 1.18 kg/m^3)

ρ_{sf} = density of the supplementary fuel (0.0408 lb/ft^3 or 0.653 kg/m^3 for CH$_4$)

T_{comb} = combustion temperature ($^\circ$F or $^\circ$C)

T_{he} = temperature of the waste gas stream after the heat exchanger ($^\circ$F or $^\circ$C)

T_{ref} = reference temperature (77 $^\circ$F or 25 $^\circ$C)

C_p = average specific heat of air between T_{comb} and T_{ref} (Btu/lb-$^\circ$F or kJ/kg-K); the specific heat capacity value can be extrapolated from data in Table 2.2.

H_{wg} = heat content of the waste gas stream (Btu/lb, kJ/kg)

H_{sf} = heating value of the supplementary fuel (21,600 Btu/lb or 50,200 kJ/kg)

If the temperature of the waste gas stream after the heat exchanger (T_{he}) is not specified, use the following equation to calculate T_{he}:

$$T_{he} = \left(\frac{HR}{100}\right)T_{comb} + \left[1-\frac{HR}{100}\right]T_{wg}$$ (7.31)

where HR = heat recovery in the heat exchanger, % (If no other information is available, a value of 70% may be assumed). If no heat exchangers are employed to recuperate the heat, then $T_{he} = T_{wg}$.

Example 7.14: Heat recovery efficiency of a heat exchanger

Referring to the recuperative heat exchanger in Figure 7.12, estimate the heat recovery efficiency of that heat exchanger.

Solution:

As shown, $T_{wg} = 120\ ^\circ$C, $T_{comb} = 700\ ^\circ$C, and $T_{he} = 351\ ^\circ$C, so

$$624 = \left(\frac{HR}{100}\right)(973) + \left[1-\frac{HR}{100}\right](393)$$

$$HR = 39.8$$

Discussion: The heat recovery efficiency of this system is 39.8%, which is less than 70%. Using temperatures in K or in $^\circ$C will achieve the same result.

Example 7.15: Supplemental fuel requirement

Referring to the remediation project described in Example 7.10, an off-gas stream ($Q = 210$ scfm $= 6$ standard m^3/min) containing 800 ppm of xylene is to be treated by a thermal oxidizer with a recuperative heat exchanger. The combustion temperature is set at 982 °C (1,800 °F). Determine the flow rate of methane as the supplementary fuel, if required.

Solution:

(a) Assuming heat recovery = 70% and T of the waste gas stream = 25 °C (77 °F), T_{he} can be found as:

$$T_{he} = \left(\frac{70}{100}\right)(982) + \left[1 - \frac{70}{100}\right](25) = 695 \ °C = (1,283 \ °F)$$

(b) The average specific heat can be derived from Table 2.2 as 1.07 kJ/kg-K (0.256 Btu/lb-°F) for the interval between 25 and 982 °C (77 to 1,800 °F)

(c) The heat content of the waste gas is 129 kJ/kg (55.6 Btu/lb), as determined in Example 7.10.

(d) The supplementary fuel required can be estimated by using Eq. 7.29 as:

$$Q_{sf} = \frac{\rho_{wg}Q_{wg}[C_p(1.1T_{com}-T_{he}-0.1T_{ref})-H_{wg}]}{\rho_{sf}[H_{sf}-1.1C_p(T_{comb}-T_{ref})]} =$$

$$\frac{(1.18)(6)\{(1.07)[1.1(982)-695-0.1(25)]-129\}}{0.653[50,200-1.1(1.07)(982-25)]} = 0.062 \ m^3/min \ (= 2.2 \ scfm)$$

The flow rate of the influent (Q_{inf}) to the combustion chamber is the sum of those of the waste gas (Q_{wg}), auxiliary (dilution or supplemental) air (Q_{aux}), and the supplementary fuel (Q_{sf}) as:

$$Q_{inf} = Q_{wg} + Q_{aux} + Q_{sf} \qquad (7.31)$$

In most cases, one can assume that the flow rate of the combined gas stream (Q_{inf}) entering the combustion chamber under the standard conditions is approximately equal to that of the flue gas leaving the combustion chamber (Q_{fg}) under the standard conditions. The change of volume across the incineration chamber, due to combustion of COCs and the supplementary fuel, is insignificant.

The flue gas flow rate under actual conditions ($Q_{fg,actual}$) can be determined from that under the standard conditions:

$$Q_{fg,actual} = Q_{fg,standard} \left[\frac{T_{comb}}{T_{standard}}\right] \qquad (7.32)$$

Just a reminder, the combustion temperature and the standard temperature in Eq. 7.32 should be the absolute temperatures.

The volume of the combustion chamber (V_{comb}) can be readily determined from the residence time, τ (in sec), and $Q_{fg,actual}$ by:

$$V_{comb} = \left[\left(\frac{Q_{fg,actual}}{60}\right)\tau\right] \times 1.05 \qquad (7.33)$$

The equation is nothing but "residence time = volume ÷ flow rate". The factor of 1.05 is a safety factor, which is an industrial practice to account for minor fluctuations in the flow rate. Table 7.4 tabulates typical thermal oxidizer system design values for toxic air pollutants.

Table 7.4 - Typical design values for thermal oxidizers (EPA, 1991)

Required Destruction Efficiency (%)	Non-halogenated Compounds		Halogenated Compounds	
	Combustion Temperature	Residence Time (sec)	Combustion Temperature	Residence Time (sec)
98	1,600 °F (538 °C)	0.75	2,000 °F (1,093 °C)	1.0
99	1,800 °F (982 °C)	0.75	2,200 °F (1,204 °C)	1.0

Example 7.16: Sizing a thermal oxidizer

Referring to the remediation project described in Example 7.15, an off-gas stream ($Q = 210$ scfm = 6 standard m^3/min) containing 800 ppm of xylene is to be treated by a thermal oxidizer with a recuperative heat exchanger. The combustion temperature is set at 982 °C (1,800 °F) to achieve a destruction efficiency of 99%. Determine the size of the thermal oxidizer.

Solution:

(a) The flue gas flow rate under standard conditions:

$Q_{fg} \sim Q_{inf} = Q_{wg} + Q_{aux} + Q_{sf} = 6.0 + 0 + 0.062 = 6.06$ m^3/min (= 214 scfm)

(b) The flue gas flow rate at actual conditions:

$Q_{fg,actual} = 6.06 \times [(273 + 982)/(273 + 25)] = 25.5$ m^3/min (= 900 ft^3/min)

(d) The required residence time is one second for 99% destruction (Table 7.4):

$$V_{comb} = \left[\left(\frac{25.3}{60}\right)(1)\right] \times 1.05 = 0.44 \; m^3 = 15.6 \; ft^3$$

Thermal oxidation of halogenated organic compounds would generate acidic gases such as HCl and HF. This type of compounds is corrosive and discharge limits are often imposed. Consequently, the emission rates of these compounds need to be calculated to determine if a removal system, such as an absorber, is needed after the thermal oxidizer.

Example 7.17: Emission rate and concentration of acid gas

A thermal oxidizer is designed to treat a 100 standard m³/min (3,650 scfm) gas stream containing 500 ppm benzene (C_6H_6) and 200 ppm carbon tetrachloride (CCl_4). The oxidizer uses 6 and 20 standard m³/min (219 and 706 scfm) of natural gas and supplemental air, respectively. Estimate the emission rate and concentration of HCl in the flue gas.

Solution:

(a) MW of CCl_4 = (12)(1) + (35.5)(4) = 154

Mass concentration of CCl_4 in the gas stream

$$= (500) \times (154/24.46) = 3,142 \; mg/m^3 = 3.14 \; g/m^3$$

Mass flow rate of CCl_4 in the influent gas stream

$$= (3.14 \; g/m^3)(100 \; m^3/min) = 314 \; g/min$$

Molar flow rate of CCl_4 in the influent gas stream

$$= 314 \; g/min \div 154 \; g/mole = 2.04 \; mole/min$$

(b) Since one mole of CCl_4 will generate four moles of HCl,

molar flow rate of HCl in the flue gas = (2.04 mole/min) × (4/1)

$$= 8.16 \; mole/min$$

MW of HCl = (1)(1) + (35.5)(1) = 36.5

Mass flow rate of HCl in the flue gas

$$= (8.16 \; mole/min)(36.5 \; g/mole) = 298 \; g/min$$

$$= 17.9 \; kg/hr \; (= 39 \; lb/hr)$$

(c) The flue gas flow rate under standard conditions:

$Q_{fg} \sim Q_{inf} = Q_{wg} + Q_{aux} + Q_{sf} = 100 + 6 + 20 = 126$ m^3/min (= 4,470 scfm)

Mass concentration of HCl in the flue gas under the standard conditions

$= (298$ g/min$) \div (126$ m^3/min$) = 2.365$ g/m$^3 = 2,365$ mg/m^3

Volume concentration of HCl in the flue gas

$= (2,365) \div [(36.4)/(24.05)] = \underline{1,560}$ ppm

7.5 Catalytic Incineration

Similar to direct thermal oxidation, catalytic incineration also employs elevated temperatures to destruct combustible COCs in gas streams. However, its combustion temperatures are lower than those of thermal oxidation for a given destruction efficiency, due to participation of catalysts in the reaction. The lower combustion temperature means less fuel usage.

7.5.1 Principles of catalytic incineration
A catalyst is a substance that increases the rate of a chemical reaction, by lowering its activation energy, but without going through permanent chemical changes itself. Typical catalysts used in catalytic incineration are precious metal oxides of platinum (Pt), palladium (Pd), or rhodium (Rh). Some base metal oxides such as vanadium pentaoxide (V_2O_5), titanium dioxide (TiO_2), or manganese oxide (MgO) are also being used.

7.5.2 Catalytic incineration systems and their components
A catalytic thermal oxidizer is very similar to a thermal oxidizer, except a catalyst bed is incorporated. Passing through a zone with burners, the waste gas stream is heated to the required temperature and then pass a bed containing the catalyst. The catalysts are deposited as a thin layer on a material with a high specific surface area, such as alumina, that is bonded to a support structure. Typical configurations of support structures are honeycombs, grids, and mesh pads so that they can provide large surface areas for reactions and facilitate uniform flow with low pressure drop.

A recuperative or regenerative heat exchanger is often employed. Figure 7.15 illustrates a recuperative catalytic thermal oxidizer.

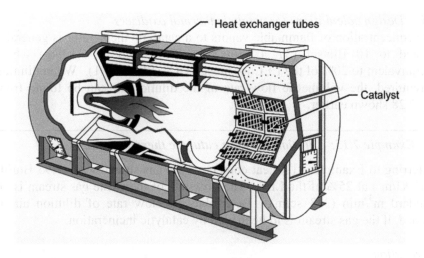

Figure 7.15 - Recuperative catalytic oxidizer (EPA, 2003)

7.5.3 Factors affecting catalytic incineration

For catalytic oxidation, the three T's (temperature, residence time, and turbulence) are still the important design parameters. In addition, type of the catalyst has a significant effect on the system performance and cost.

Most organic COCs have catalytic auto-ignition temperatures of about 400 to 500 °F (~ 200 to 260 °C). It should be noted that they are smaller than their corresponding auto-ignition temperatures (see Table 7.3). The inlet temperature entering the catalyst bed is generally 100 °F (55.5 °C) above the catalytic auto-ignition temperature of the most difficult-to-destroy COC in the gas stream, so that the minimum inlet temperate to the bed is around 500 to 600 °F (260 to 315 °C).

The temperature of the gas stream would increase about 27 °F (15 °C) for each 1% of the LEL oxidized in the catalyst bed. For a gas stream at its 25% LEL, its temperature can go up by 675 °F (375 °C) which will bring the temperatures of the catalyst bed and the gas stream at its outlet to about 1,175 to 1,275 °F (635 to 690 °C). This is very close to the temperature that can deactivate/damage most catalysts (LADCO, 2009). This is the reason why the organic vapor concentrations entering catalytic thermal oxidizers are usually kept less than 20 or 25% of the LEL.

7.5.4 Design calculations for catalytic thermal oxidizers

The concentration of flammable vapors to a catalytic incinerator is generally limited to 10 Btu/scf (372.4 kJ/m^3) or 135 Btu/lb (314 kJ/kg) which is equivalent to 20% of the LEL for most VOCs (EPA, 1991). When dilution is required, the volumetric flow rate of the dilution air can be found from Eq. 7.28 shown earlier.

Example 7.15: Dilution air for a catalytic thermal oxidizer

Referring to Example 7.11, heat content of the gas stream was 11.95 Btu/ft^3 (445 kJ/m^3) at 25% of the LEL. The flow rate of the waste gas stream is 10 standard m^3/min (353 scfm). Determine the flow rate of dilution air, if needed, if the gas stream is to be treated by catalytic incineration.

Solution:

The heating content of the waste gas stream at 25% of its LEL exceeds the 10 Btu/scf (372.4 kJ/m^3) limit. Therefore, air dilution is required, and the flow rate of the dilution air can be found as:

$$Q_{dilution} = \left[\left(11.95/10\right) - 1\right](10) = 1.95 \ m^3/min \ (= 65.3 \ scfm)$$

Discussion: For this waste gas stream, dilution air may not be needed for thermal oxidation, but it would be needed for catalytic incineration.

For catalytic incineration, supplementary heat can be provided by electrical heaters. If natural gas is used, use Eq. 7.29 to determine the required flow rate of the supplementary fuel flow rate. Before calculating the supplementary heat requirement, the following two equations should be used to estimate the temperature of the flue gas, T_{out}, which can achieve the desired destruction efficiency without damaging the catalyst. The T_{out} can be estimated with the temperature of the waste gas after the heat exchanger but before the catalyst bed (T_{in}) and the heat content of the gas:

$$T_{out} \ (^oF) = T_{in}(^oF) + 50 \times H_w (in \ Btu/scf) \qquad (7.34)$$

$$T_{out} \ (^oC) = T_{in}(^oC) + 0.746 \times H_w (in \ kJ/m^3) \qquad (7.35)$$

where H_w is the heat content of the waste gas stream.

These two equations assume a 50 °F temperature increase for every Btu/scf of heat content (or 0.746 °C temperature increase for every kJ/m^3) in the influent to the catalyst bed. These equations can also be used to determine the required influent temperature to achieve a desired temperature in the catalyst bed.

Example 7.16: Dilution air for a catalytic thermal oxidizer

Referring to the remediation project described in Example 7.13, an off-gas stream (Q = 6 m³/min = 210 scfm) containing 800 ppm of xylenes is to be treated by a catalytic incinerator with a recuperative heat exchanger. After the heat exchanger, the temperature of the diluted waste gas is 550 °F (288 °C). Estimate the temperature of the catalyst bed.

Solution:

After air dilution, heat content of the diluted waste gas is 10 Btu/scf (372.4 kJ/m³).

$$T_{out} = 550 + 50 \times 10 = 1,050 \ ^oF; \text{ or}$$

$$T_{out} = 288 + 0.746 \times 372.4 = 566 \ ^oC$$

Discussion: The calculated temperature, 1,050 °F (566 °C), falls within the typical temperature range for catalyst beds (1,000 - 1,200 °F, or 540 - 650 °C).

Similar to thermal oxidation, the flow rate of the influent to a catalyst bed is the sum of those of the waste gas, auxiliary (dilution or supplemental) air and the supplementary fuel. Because of the short residence time in the catalyst bed, space velocity (SV) is commonly used to relate the volumetric influent flow rate and the volume of the catalyst bed. The *space velocity* is defined as the volumetric flow rate of the influent entering the catalyst bed divided by the volume of the catalyst bed. It is actually the inverse of the residence time. Table 7.5 provides the typical design parameters for catalytic incinerators. It should be noted here that the flow rate used in the space velocity calculation is the influent gas flow rate at standard conditions, not that of the catalyst bed or the bed effluent.

Table 7.5 - Typical Design Parameters for Catalytic Thermal Oxidizers

Desired Destruction Efficiency (%)	T at the Bed Inlet	T at the Bed Outlet	Space Velocity (hr⁻¹)	
			Base Metal	Precious Metal
95	600 °F (315 °C)	1,000 - 1,200 °F (540 - 650 °C)	10,000 - 15,000	30,000 - 40,000

The size of the catalyst can be determined by:

$$V_{cat} = \frac{60 \times Q_{inf}}{SV} \qquad (7.36)$$

where

V_{cat} = volume of the catalyst bed (in m^3 or ft^3)

Q_{inf} = influent flow rate to the catalyst bed, in (standard m^3/min or scfm)

SV = space velocity, hr^{-1}

Example 7.17: Sizing a catalytic thermal oxidizer

Referring to the remediation project described in Example 7.15, an off-gas stream (Q = 210 scfm = 6 standard m^3/min) containing 800 ppm of xylene is to be treated by a catalytic thermal oxidizer with a recuperative heat exchanger. The design space velocity is 12,000 hr^{-1}. Determine the size of the catalyst bed.

Solution:

(a) The flue gas flow rate at standard conditions:

$Q_{fg} \sim Q_{inf} = Q_{wg} + Q_{aux} + Q_{sf} = 6.0 + 0 + 1.95 = 7.95$ m^3/min (= 281 scfm)

(b) With a space velocity of 12,000 hr^{-1}, the size of the catalyst bed:

$$V_{cat} = \frac{60 \times 7.95}{12,000} = 0.04 \; m^3 \; (= 1.4 \; ft^3)$$

Discussion: The size of the catalyst, 0.04 m^3 (1.4 ft^3), is smaller than the volume of the combustion chamber needed for thermal oxidation, 0.44 m^3 (15.6 ft^3).

7.6 Condensers

Using condensation for air pollution control, COCs are removed from the gas stream by causing them to change from gaseous state to liquid state. It can be done by reducing the system temperature, increasing the system pressure, or a combination of both. However, most condensers for air pollution control use temperature reduction because it is less expensive than compressing a gas stream. When compared to other air pollution control technologies, condensation is more favorable for the cases of high COC concentrations and the condensed products having good recycle/reuse potentials.

7.6.1 Principles of condensation

As a COC-laden gas stream is cooled down, the partial pressure of a COC in the gas stream will decrease. When the partial pressure drops to its vapor pressure, the condensation will start. With further reduction in the system temperature, the vapor pressure of the COC in the liquid state decreases which means that its concentration in the gas stream gets lower.

A coolant is needed to absorb the heat/energy from the gas stream and, in turn, its temperature will increase. It is essentially a heat-transfer process. The condensation process is typically applied to the cases where the organic COCs can be economically recovered. It is also limited to the cases where only one or two organic compounds in the gas stream to be recovered because separation of multiple organic compounds in the condensate could be difficult. In addition, particulates in gas streams can cause fouling problems in most of the heat transfer systems, this process is typically limited to gas streams with low particulate concentrations (EPA, 2002).

7.6.2 Condensation systems and their components

A condensation control system needs to facilitate the heat transfer between the gas stream and the coolant. Three types of condensers are used for air pollution control: conventional, refrigeration, and cryogenic systems.

Conventional condensation systems are relatively simple in that they employ a coolant (usually chilled water) to reduce the temperature of the gas stream to as low as 40 °F (4.4 °C). Using brine as the coolant can reduce the gas stream to 0 °F (-17.8 °C). There are two types of conventional systems; one uses direct-contact condensers and the other uses surface condensers. A *direct-contact condenser* is physically similar to an absorber/wet scrubber, except that the liquid has a much lower temperature to condense the COC in the gas stream in addition to absorption, if the COC is water soluble. The coolant and the condensed/absorbed COC leave the condenser unit as a single stream that would need further treatment. The condenser units can be spray towers, tray towers, or ejectors (Figure 7.16). Ejectors employ high pressure spray nozzles to create suction that moves the inlet gas stream. The coolant is sprayed into the throat of the ejector and COC vapors will condense while passing through. Direct-contact condensers are simple and inexpensive. However, a wastewater stream containing the spent coolant and COCs often needs to be treated.

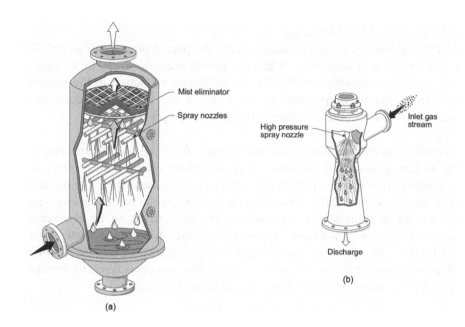

Figure 7.16 - Direct-contact condensers: (a) spray tower and (b) ejector (EPA, 2003)

A typical *surface condenser* is a shell-and-tube heat exchanger, similar to the exchanger used in a recuperative thermal oxidizer (Figure 7.13). The coolant flow through the tubes while the waste gas flows through the shell outside the tubes. The COCs will condense on the outside surface of the tubes and then collected. The tubes are made of materials that are conductive to heat, such as copper.

Refrigeration systems have lower operating temperatures than those of the conventional system by using refrigerants and compressors. As shown in Figure 7.17, the COC-laden gas stream is drawn into a pre-condenser where it cooled to ~ 40 $^{\circ}$F (4.4 $^{\circ}$C) to reduce water content to <0.02 lb H_2O/lb air (kg H_2O/kg air). It is to minimize build-up of frost on top of the heat exchange surfaces of the main chamber which operates at much lower temperatures, ranging from -50 to -150 $^{\circ}$F (-43 to -101 $^{\circ}$C) (EPA, 2003).

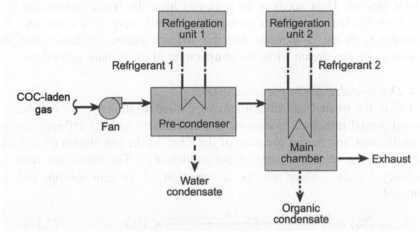

Figure 7.17 - Process flow diagram of a refrigeration system (modified from EPA, 2003)

Cryogenic systems use liquefied nitrogen or carbon dioxide, with operating temperatures ranging from -100 to -320 °F (-73 to -160 °C) to cool waste gas streams to the freezing point of the COCs (EPA, 2003). Figure 7.18 illustrates a simplified process flow diagram of a cryogenic system.

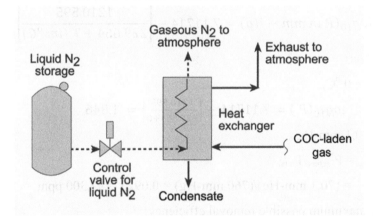

Figure 7.18 - Process flow diagram of a cryogenic system (modified from EPA, 2003)

7.6.3 Factors affecting condensation
Effectiveness of a condensation system depends on the characteristics of the COCs and the coolant, configuration of the system, and operation parameters. As mentioned, condensation for air pollution control is essentially a heat

241

transfer process. Heat needs to be removed from the waste stream and it is picked up by the coolant. Consequently, condensate is formed and the temperatures of the condensate and the effluent stream are lower than the influent waste gas stream while the temperature of the coolant will increase.

7.6.4 Design calculations for condensers

A COC in the treated gas stream leaving a condensation system will have a reduced partial pressure. Assuming the condensate and the effluent gas are in equilibrium, the partial pressure of the COC in the gas stream (P_{COC}) will be equal to the vapor pressure of the condensate. The maximum removal efficiency (η_{max}) , based on the assumption of an equilibrium, can be estimated by:

$$\eta_{max} \ (\%) = \frac{P_{COC, \ influent \ gas} - VP_{condensate}}{P_{COC, \ influent \ gas}} \times 100 \qquad (7.37)$$

Example 7.18: Removal efficiency of a condenser

What would be the maximum possible removal efficiency for acetone (C_3H_6O) in an indirect contact condenser operating at 0 °C (32 °F) at 1 atm? The acetone concentration in the influent gas stream is 500,000 ppm. The Antoine equation for acetone (T = -12.9 to 55.3 °C) is shown below (EPA, 2012):

$$log_{10}(P \ in \ mm - Hg) = 7.11714 - \left[\frac{1210.595}{229.664 + T \ (in \ °C)}\right]$$

Solution:

(a) At T = 0 °C,

$$log_{10}(P) = 7.11714 - \left[\frac{1210.595}{229.664+0}\right] = 1.846$$

P = 70.1 mm-Hg

$y_{acetone} = P^{*}_{actone}/P_{total}$

$= (70.1 \ mm-Hg)/(760 \ mm-Hg) = 0.0923 = 92,300 \ ppm$

(b) The maximum possible removal efficiency:

$$\eta_{max}(\%) = \frac{300,000 - 92,300}{300,000} \times 100 = 69.2$$

Discussion: Vapor concentrations in ppm, instead of partial pressure, were used to determine the maximum removal efficiency.

Often the influent gas stream is above its dew point (superheated), it needs to be cooled to its dew point so that condensation can begin. In addition, the condensation needs to be sub-cooled so that it will not re-vaporize. This is typically done by maintaining a condensate level that covers the tubes in the bottom of the last heat exchanger, a condition known as *flooding*. The total heat needs to be taken away by the coolant include (i) the sensible heat change of the waste gas, (ii) the latent heat (heat of condensation or vaporization) of the condensate, and (iii) the sensible heat change of the condensate from the dew point to the after sub-cooling temperature. The total rate of heat transfer (q) can be determined as:

$$q = (m_g)(C_{p,g})(T_{g,i} - T_{g,f}) + (m_{con})(\Delta H_v)$$

$$= (m_L)(C_{p,L})(T_{L,f} - T_{L,i}) \quad (7.38)$$

where m_g, m_{con}, and m_L are the mass flow rate of the gas, the condensate and the coolant; $C_{p,g}$ and $C_{p,L}$ are the specific heat capacities of the gas and the liquid; $T_{g,i}$, $T_{g,f}$, $T_{L,i}$ and $T_{L,f}$ are the initial and final temperatures of the gas and the liquid; and ΔH_v is the heat of vaporization of the condensate.

For an indirect condenser, the rate of heat transfer across the tube surface depends on overall heat transfer coefficient (U), the heat transfer surface area (A), and the temperature difference between the gas and liquid across the tube (ΔT) as:

$$q = U \times A \times \Delta T \quad (7.39)$$

The overall heat transfer coefficient (U) is a measure of the total resistance to heat transfer from the hot side to the cool side across the tube. Typical values of universal heat transfer constant, for the cases where the gas stream is organic solvent vapor with high percentages of non-condensable gases and water is the coolant, are 20 to 60 Btu/°F-ft^2-hr, or 410 to 1,230 kJ/hr-m^2-K (EPA, 2012). [Note that 1 Btu/°F-ft^2-hr = 20.44 kJ/hr-m^2-K = 0.34 kJ/min-m^2-K].

In an indirect contact heat exchanger, the gas stream and the coolant typically flow concurrently or counter-currently. For either case, the temperature difference between the two fluids varies throughout the entire length of the heat exchanger. To account for the variations, the long mean of the temperature differences ($\Delta T_{log\ mean}$) is usually used and it can be found by:

$$\Delta T_{log\ min} = \frac{(\Delta T_{inlet} - \Delta T_{outlet})}{ln\left(\Delta T_{inlet} / \Delta T_{outlet}\right)} \quad (7.40)$$

243

In almost all condensation applications used for air pollution control, the waste gas to be treated will contain one or more condensable COCs with non-condensable components. The non-condensable components would offer resistance to both heat and mass transfer and this needs to be taken into consideration in design. The following two examples are for a simplified case (pure solvent) and they were adapted from EPA (2012). The main purpose is to illustrate calculations related to heat transfer.

Example 7.19: Required mass flow rate of coolant for condensation

A single-pass, counter-current flow, indirect-contact condenser is used for a gas stream of pure acetone (C_3H_6O; MW = 58.08) at T = 120 °C and P =1 atm. The boiling point of acetone is 56.2 °C (133.2 °F). The feed rate of acetone is 600 kg/hr. Water is the coolant that enters the condenser at 20 °C and exits at 40 °C. Using the following information to determine the coolant flow rate:

- Heat capacity of acetone vapor = 0.084 kJ/gmole-K
- Heat capacity of acetone liquid = 0.13 kJ/gmole-K
- Heat capacity of water = 4.182 kJ/kg-K = 0.0753 kJ/gmole-K
- Heat of vaporization of acetone = 25.1 kJ/gmole-K

Solution:

(a) Since the given heat capacity values are on the molar basis,

Molar flow rate of acetone = (600 kg/hr)(1,000 g/kg) ÷ (58.08 g/g-mole)

= 10,330 g-mole/hr = 172 g-mole/min

(b) Total heat needs to be removed (q) = (cooling the superheat vapor from 120 °C to its boiling point of 56.2 °C) + (isothermal condensation of acetone at 56.2 °C) + (sub-cooling the condensate from 56.2 to 30 °C)

$q = (m_g)(C_{p,g})(T_{g,i} - T_{g,f}) + (m_{con})(\Delta H_v) = (172)\{(0.084)(120 - 56.2) + 25.1 + (0.13)(56.2 - 30)\} = 5,820 \; kJ/min$

(c) Total heat taken away by the coolant (q):

$$q = 5,820 \; kJ/min = (m_L)(C_{p,L})(T_{L,f} - T_{L,i})$$

$$= (m_L)(4.182)(40 - 20)$$

Coolant flow rate (m_L) = 69.6 kg/min

Discussion: We assumed that all the heat capacity values are constant. It is justifiable because all the temperature ranges are relatively narrow.

Example 7.20: Surface area of heat exchangers for condensation

Referring to Example 7.19, use the overall heat transfer coefficient (U) values given below to determine the required surface area of heat exchangers:

- $U = 40$ Btu/$^\circ$F-ft^2-hr (cooling superheated acetone vapor)
- $U = 100$ Btu/$^\circ$F-ft^2-hr (condensing acetone)
- $U = 50$ Btu/$^\circ$F-ft^2-hr (sub-cooling acetone liquid)

[Note: 1 Btu/$^\circ$F-ft^2-hr = 20.44 kJ/hr-m^2-K = 0.34 kJ/min-m^2-K]

Solution:

(a) For the stage of cooling superheated acetone vapor:

Removal of superheat (q) = $(m_g)(C_{p,g})(\Delta T)$

$$= (172)(0.084)(120 - 56.2) = 922 \text{ kJ/min}$$

Temperature change of the coolant $(\Delta T) = q \div [(m_L)(C_{p,L})]$

$$= 922 \div [(69.6)(4.182)] = 3.2 \text{ K}$$

Temperature of the coolant at the end of this stage = $40 - 3.2 = 36.8$ $^\circ$C

$$\Delta T_{\log \min} = \frac{(\Delta T_{inlet} - \Delta T_{outlet})}{\ln\left(\Delta T_{inlet}/\Delta T_{outlet}\right)} = \frac{[(120 - 40) - (56.2 - 36.8]}{\ln\left[(120 - 40)/(56.2 - 36.8)\right]}$$

$$= 42.8$$

Required heat transfer area $(A_1) = q \div [(U)(\Delta T_{\log \text{ mean}})]$

$$= 922 \div \{[(40)(0.34)](42.8)\} = \underline{1.6 \text{ m}^2}$$

(b) For the stage of acetone condensation:

Heat of condensation (q) = $(m_{con})(\Delta H_v) = (172)(25.1) = 4{,}320 \text{ kJ/min}$

Temperature change of the coolant $(\Delta T) = q \div [(m_L)(C_{p,L})]$

$$= 4{,}320 \div [(69.6)(4.182)] = 14.8 \text{ K}$$

Temperature of the coolant at the end of this stage = 36.8 – 14.8 = 22.0 °C

$$\Delta T_{\log min} = \frac{(\Delta T_{inlet} - \Delta T_{outlet})}{\ln \left(\Delta T_{inlet} / \Delta T_{outlet} \right)} = \frac{[(56.2 - 36.8) - (56.2 - 22]}{\ln \left[(56.2 - 36.8) / (56.2 - 22) \right]}$$

$$= 26.1 \ K$$

Required heat transfer area (A_2) = q ÷ $[(U)(\Delta T_{\log mean})]$

= 4,320 ÷ {[(100)(0.34)](26.1)}= $\underline{4.9 \ m^2}$

(c) For the stage of sub-cooling condensate:

Removal of superheat (q) = $(m_{con})(C_{p,con})(\Delta T)$

= (172)(0.13)(56.2 - 30) = 586 kJ/min

Temperature change of the coolant (ΔT) = q ÷ $[(m_L)(C_{p,L})]$

= 586 ÷ [(69.6)(4.182)] = 2.0 K

Temperature of the coolant at the end of this stage = 22.0 – 2.0 = 20.0 °C

$$\Delta T_{\log min} = \frac{(\Delta T_{inlet} - \Delta T_{outlet})}{\ln \left(\Delta T_{inlet} / \Delta T_{outlet} \right)} = \frac{[(56.2 - 22.0) - (30 - 20]}{\ln \left[(56.2 - 22.0) / (30 - 20) \right]} = 19.7$$

Required heat transfer area (A_3) = q ÷ $[(U)(\Delta T_{\log mean})]$

= 586 ÷ {[(50)(0.34)](19.7)}= $\underline{1.7 \ m^2}$

(d) Overall temperature change of the coolant = 3.2 + 14.8 + 2.0 = 20 K (check)

(e) The total surface area = 1.6 + 4.9 + 1.7 = 8.2 m^2

Discussion: The area needed for the second stage (condensation of acetone) is much larger than those of the other two stages for this case.

7.7 Bioreactors

Bioreactors use microorganisms to degrade COCs in waste gas streams. Although bioreactors can be aerobic or anaerobic, aerobic is usually the mode. They are mainly for waste gas streams with relatively low concentrations of COCs, which are soluble in water and biodegradable. Biodegradation is especially applicable to most of the organic COCs and some inorganic COCs such as hydrogen sulfide (H_2S) and nitrous oxide (N_2O). Bioreactors are also commonly used for odor control in wastewater treatment plants. Removal efficiencies range from 60% to >99% (EPA, 2002).

7.7.1 Principles of bio-reaction

Aerobic biodegradation of hydrocarbons is an oxidation process, in which microorganisms (M/O) use oxygen as an electron acceptor to degrade organics (the electron donors) for cell synthesis and energy. The final products will be CO_2 and H_2O, same as those from thermal oxidation, except heat release from thermal oxidation while more biomass will be generated in biodegradation. The other major difference is that biodegradation typically occur under ambient temperatures.

$$C_xH_y + \left(x + \frac{y}{4}\right)O_2 \xrightarrow{M/O} x\,CO_2 + \left(\frac{y}{2}\right)H_2O + Biomass \qquad (7.41)$$

7.7.2 Bioreactors and their components

There are three main types of bioreactors for air pollution control: biofilters, biotrickling filters, and bioscrubbers.

A *biofilter* can be as simple as a box which contains a plenum (an enclosed space used for air flow), a support rack above the plenum, and bed media on top of the support media (Figure 7.19). Various materials are used for bed media such as peat, compost, bark, and lava rock. The bed media are to allow M/O to form a biofilm where the biodegradation would take place. Typical beds are 3 to 5 feet (1 to 1.5 m) deep with a porosity of ~50% and 60% of water saturation. Water with nutrients can be irrigated on top of the biofilter bed or the COC-laden gas will be humidified before being introduced into the bio-filter. As the waste gas stream flows through the bed media, the COCs are absorbed by the moisture in the biofilm and come into contact with M/O and get biodegraded. Typical operating temperatures of the bed are around 60 to 85 °F (15 to 30 °C) and the pH of the support media is typically 6 to 8. Residence time of the waste gas stream in the bed media ranges from 15 to 60 minutes (EPA, 2002). COCs containing sulfur or nitrogen may generate acidic byproducts during biodegradation which may require alkali to keep the pH in the desirable range. Sometimes oyster shells are used as part of the bed media for neutralizing acid build-up or spraying dilute soda ash (Na_2CO_3) over the top of the media bed intermittently (EPA, 2003b).

247

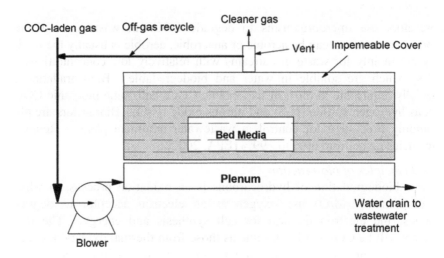

**Figure 7.19 - A biofilter system with off-gas recycle
(modified from EPA, 2003b)**

The biofilter in Figure 7.19 is equipped with a cover and a vent which allows recycling of the gas stream to enhance the removal efficiency. The biofilters can also be placed in the ground (Figure 7.20) which are common in wastewater treatment plants for odor control. They may appear as planters.

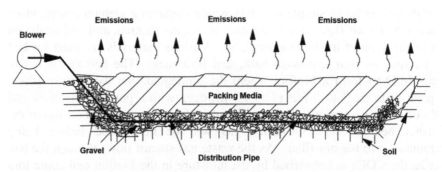

**Figure 7.20 - An in-ground biofilter system
(modified from EPA, 2003b)**

Figure 7.21 illustrates the process flow diagram of a biofilter which includes pretreatment for particulate, temperature, and load control. The influent waste gas stream is humidified; additional water, nutrient, and pH buffer are intermittently added to achieve optimal conditions for M/O activities. Leachate collection may also be needed.

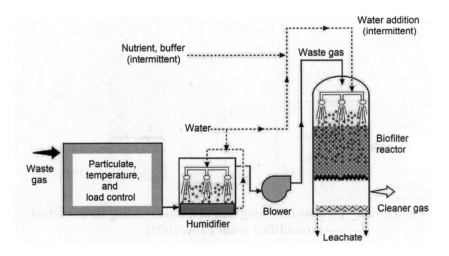

**Figure 7.21 - Process flow diagram of a biofilter system
(modified from EPA, 2002)**

A biotrickling filter operates under the same principle as a trickling filter of a wastewater treatment. A trickling filter in wastewater treatment contains media, made of aggregate, ceramic, or plastic, to support the growth of biofilm. A wastewater stream trickles down from the top and the COCs in it get absorbed into the liquid in the biofilm, while clean air is induced through the bed and provides the oxygen to the M/O in the biofilm for biodegradation. For a biotrickling filter in air pollution control, the COCs to be removed are in the waste gas stream, not in the water. When compared to biofiltration for air pollution control, water is recirculated and nutrients and acids/bases can be added to maintain an environment conducive inside the biotrickling filter for optimal COC removal. It is less prone to clogging by biomass and to build-up of acid, but a bio-trickling would produce a wastewater stream that may requires further treatment (Figure 7.22). Use of ceramic or plastic media can achieve a porosity of up to 95%, which should greatly reduce the pressure drop of the gas flow through the media. For a same depth of bed media, the pressure drop of a biotricking filter can be as low as one-fifth of that of a biofilter with natural packing materials (EPA, 2003b).

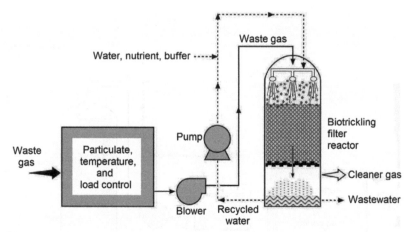

Figure 7.22 - Process flow diagram of a bio-trickling filter system (modified from EPA, 2002)

As discussed, a bio-tricking filter is considered as an enhancement of a biofilter; a bioscrubber can be considered as an enhancement of a biotrickling filter. When compared to biotrickling filters, the bioscrubbers improve the absorption of COCs in the liquid and lengthen the time for M/O to degrade the COCs. Absorption takes place in a typical absorber/wet scrubber such as spray towers, plate towers, or packed columns. The effluent from the wet scrubber is then transferred to an aeration tank in which the COCs are degraded by suspended M/O. Water, nutrients, and buffer solutions can be added to provide optimal conditions for microbial activities. Solution in the aeration tank can be recycled back to the scrubber (Figure 7.22). The potential for bio-clogging is minimal for bio-trickling filters.

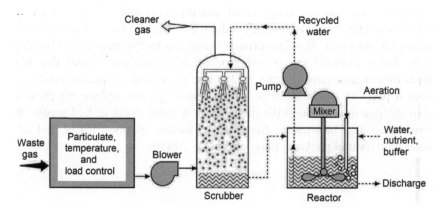

Figure 7.22 - Process flow diagram of a bioscrubber system (modified from EPA, 2002)

7.7.3 Factors affecting performance of bioreactors

Key parameters for design of a bioreactor include concentrations and biodegradability of the COC, types of the M/O present, media, residence time, presence of oxygen for aerobic biodegradation (or absence of oxygen for anaerobic biodegradation), pH, temperature, moisture, presence of available nutrients, and free of toxins.

M/O can be classified by the temperature range for their optimal growth. A *mesophile* will grow best in moderate temperatures, typically between 20 and 45 °C (68 and 113 °F) [Note: The optimal temperature range for thermophilic M/O is 45 to 122 °C (113 to 252 °F) and that of psychrophilic M/O is -15 (or lower) and 10 °C (5 °F, or lower, and 50 °F)]. Most of the bioreactors count on mesophilic M/O to do the work. If the temperature of the waste gas stream is outside this range, it needs to be adjusted before entering the bioreactor. Humidification of the gas stream by water would lower its temperature, in addition to providing the required moisture. There are also optimal pHs for bacterial growth. The M/O active in most bioreactors prefer neutral or near the neutral pH. Chemicals to adjust or buffer the pH are often added.

Aerobic M/O cannot grow without an ample supply of oxygen while anaerobic live in an environment absent of oxygen. Since the rates of biodegradation of hydrocarbons under aerobic conditions are generally higher than those under anaerobic conditions, most bioreactors are in the aerobic mode. Sufficient residence time for the waste gas in the bed media is needed to achieve the desirable removal for a specific biodegradation rate.

M/O need moisture to survive and the moisture is one of the main components of the biofilm of biofilters that absorbs COCs from the gas stream. Although the higher moisture content of bed media would facilitate more absorption, but it will also reduce the passage for the gas stream. The degree of water saturation (the void occupied by water) of the media bed should be kept between 40 and 60%. To acheive optimal efficiency, the waste gas stream should be 20 to 40 °C and 95% relative humidity (EPA, 1992).

Macronutrients (i.e., N, P. K, and others) are also needed for M/O growth. They may need to be added. Some of the compounds in the air stream, including COCs, in elevated concentrations may become toxic to M/O. Their concentrations may need to be reduced before entering the bioreactors. In addition, some COCs may be more resistant to biodegradation which should be taken into consideration.

251

7.7.4 Design calculations for bioreactors

Bio-filtration is cost effective for large volume air streams with relatively low concentrations (<1,000 ppm as methane). Maximum influent organic COC concentrations have been found to be 3,000 to 5,000 mg/m^3 (1.9 to 3.0 × 10^{-4} lb/ft^3). The media bed should be maintained at 40 - 60% of degree of saturation and a pH between 7 and 8. Typical gas loading rates (GLR) to biofilters have been designed to treat 1,000 to 150,000 m^3/hr (600 to 88,000 cfm) waste gas with the systems having 10 to 2,000 m^2 (110 to 21,500 ft^2) of filter media. The typical surface loading rate is 100 m^3/hr of waste air stream per m^2 filter cross-sectional area, or 100 m/hr (5.5 ft/min) (EPA, 1992). The required cross-sectional area of the bio-filter ($A_{biofilter}$) can be determined from the influent flow rate of the gas stream (Q_{inf}) and the GLR:

$$A_{biofilter} = \frac{Q_{inf}}{GLR} \qquad (7.42)$$

Example 7.20: Sizing biofilters for off-Gas treatment

Referring to the remediation project described in Example 7.15, an off-gas stream (Q = 210 scfm = 6.0 standard m^3/min) containing 800 ppm of xylene is to be treated by biofilters. Determine the size of the biofilters needed for this project.

Solution:

(a) The off-gas contains 800 ppm of xylenes, which is equivalent to 6,400 ppm as methane (each xylenes molecule contains eight carbon atoms). This is beyond the typical range of <1,000 ppm as methane. The maximum influent VOC concentrations of 3,000 - 4,000 mg/m^3 have been reported in literature. Although the xylenes concentration in this case (800 ppm of xylenes = 3,460 mg/m^3) falls within the range, dilution of off-gas could be a conservative approach. The optimal influent concentration should be determined from a pilot study. In this example, let us dilute the off-gas four times, therefore the influent flow rate to the biofilter becomes 24 standard m^3/min (840 scfm).

(b) Using the typical surface loading rate is 100 m^3/hr/m^2, the required cross-sectional area as:

$$A_{biofilter} = \frac{Q_{inf}}{GLR} = \frac{(24\frac{m^3}{min})(60\frac{min}{hr})}{100\ m^3/_{hr}/m^2} = 14.4\ m^2 = (155\ ft^2)$$

(c) Use 1.5 m (5 ft) as the thickness of the media bed as a conservative approach without conducting a pilot study.

Discussion: If the biofilter is constructed in a cylindrical shape, its diameter would be around 4.5 m (14 ft). Having two biofilters in series or in parallel should also be considered as alternatives.

Summary

1. The gaseous air pollutants can be organic or inorganic.
2. Volatile organic compounds are often toxic and they are important players in photochemical smog formation.
3. Common processes to reduce VOC emissions from stationary sources include absorption, condensation, adsorption, biodegradation, thermal incineration, and catalytic incineration.
4. To have effective removal by adsorption, the organic pollutants need to be relatively soluble in the water or the absorbent solution. Additives can be introduced to enhance absorption.
5. To have effective removal by condensation, the organic pollutants need to be condensable at relatively high temperatures to be economical feasible. If the pollutants are condensable and reusable, the process is a good alternative for pollutants with high concentrations.
6. For biodegradation, the organic pollutants need to be readily biodegradable. The process is also practically limited to low concentration levels. For high concentration, dilution and a larger footage would be required.
7. Adsorption and two thermal processes are generally applicable to a wide range of organic pollutants. For high pollutant concentrations, thermal processes are more preferable because adsorbents have limited capacities. On-site regeneration of adsorbents can be an alternative for high pollutant concentrations. Incineration of pollutants containing sulfur and chlorine would generate acidic gases.
8. Thermal and biological processes are destruction processes, while adsorption, absorption, and condensation are just mass transfer processes. Wastes are usually generated as by-products from the non-destructive processes which may require additional treatment and/or handling.

Bibliography

de Nevers, N (2000). *Air Pollution Control Engineering (2ⁿᵈ edition)*, McGraw-Hill Companies, Inc.

Kuo, J. (2014). *Practical Design Calculations for Groundwater and Soil Remediation* (2ⁿᵈ edition), CRC Press, Boca Raton, Florida.

Kuo, J.F. and Cordery, S.A. (1988). *Discussion of Monograph for Air Stripping of VOC from Water, J. Environ. Eng.*, V. 114, No. 5, p. 1248-50.

LADCO (2009). *APTI 400: Introduction to Air Toxics - Student Manual*, Lake Michigan Air Directors Consortium (LADCO), Rosemont, Illinois 60018.

Lide, D.R. (1992). *Handbook of Chemistry and Physics*, 73rd edition, CRC Press, Boca Raton, Florida.

Reid, R.C., Prausnitz, J.M. and Poling, B. F (1987). *The Properties of Liquids and Gases*, 4th edition, McGraw-Hill, Inc., New York, NY.

USEPA (1991). *Control Technologies for Hazardous Air Pollutants*, EPA/625/6-91/014, Office of Research and Development, US Environmental Protection Agency, Washington DC 20460.

USEPA (1992). *Control of Air Emissions from Superfund Sites.* EPA/625/R-92/012, Office of Research and Development, United States Environmental Protection Agency, Washington, DC 20460.

USEPA (1994). *Control Techniques for Fugitive VOC Emissions from Chemical Process Facilities.* EPA/625/R-93/005, Office of Research and Development, United States Environmental Protection Agency, Washington, DC 20460.

USEPA (1999). *APTI 444: Air Pollution Field Enforcement - Student Manual*, Office of Air Quality Planning and Standards, United States Environmental Protection Agency, Research Triangle Park, NC 27711.

USEPA (2002). *APTI 482: Sources and Control of Volatile Organic Air Pollutants - Student Manual (3rd edition)*, prepared by J.W. Crowder for Air Pollution Training Institute, United States Environmental Protection Agency, Research Triangle Park, NC 27711.

USEPA (2002). *EPA Air Pollution Control Cost Manual (6th edition)*, EPA/452/B-02-001, Office of Air Quality Planning and Standards, United States Environmental Protection Agency, Research Triangle Park, NC 27711.

USEPA (2003). *APTI 452: Principles and Practices of Air Pollution Control - Student Manual (3rd edition)*, Air Pollution Training Institute, United States Environmental Protection Agency, Research Triangle Park, NC 27711.

USEPA (2003). *APTI 455: Inspection of Gas Control Devices and Selected Industries - Student Manual (2nd revision)*, Air Pollution Training Institute (APTI), Environmental Research Center, United States Environmental Protection Agency, Research Triangle Park, NC 27711.

USEPA (2006). *Off-Gas Treatment Technologies for Soil Vapor Extraction Systems: State of the Practice*, EPA/542/R-05/028, Office of Superfund Remediation and Technology Innovation, Office of Solid Waste and

Emergency Response, United States Environmental Protection Agency, Washington, DC 20460, March 2006.

USEPA (2012). *APTI 415: Control of Gaseous Emissions - Student Guide,* Office of Air and Radiation, United States Environmental Protection Agency, Research Triangle Park, NC 27711.

USEPA (2012). *APTI 427: Combustion Source Evaluation - Student Manual,* Office of Air and Radiation, United States Environmental Protection Agency, Research Triangle Park, NC 27711.

Exercise Questions

1. The LEL and UEL of propane (C_3H_8) are 2.05% and 11.38% by volume, respectively. There is a propane-air mixture with an A/F ratio of 20.0 kg air/kg propane. Can it be ignited?

2. The stoichiometric mixture of methane (CH_4) and air contains 9.48% methane by volume. What is the A/F ratio for combustion of this mixture (in lb/lb or kg/kg)?

3. Assuming air contains 20% by volume O_2 and 80% N_2, for combustion of 100 standard cubic meter of methane (CH_4) in air, determine:

$$CH_4 + 2O_2 \rightarrow CO_2 + 2\,H_2O$$

   (a) The stoichiometric amount of O_2 needed for a complete combustion

   (b) The stoichiometric amount of air needed for a complete combustion

   (c) The stoichiometric mixture by volume (in % methane in the mixture)

   (d) The A/F ratio (in a mass/mass ratio)

4. To burn a pure methane (CH_4) stream of 100 standard cubic meter ($T_{standard}$ = 25 °C) with 25% excess air (20% O_2 and 80% N_2 by volume). Assuming the combustion is complete and the flue gas temperature is 300 °C,

   (a) The total influent flow rate to the oxidizer (in scfm) =

   _____

   (b) The flow rate of the flue gas (in acfm) = _____

   (c) The CO_2 concentration in the flue gas (in %) = _____

5. Assuming air contains 20% by volume O_2 and 80% N_2, for combustion of 100 standard m^3/min of ethane (C_2H_6) in air, determine

$$C_2H_6 + 3.5\,O_2 \rightarrow 2CO_2 + 3H_2O$$

   (a) The stoichiometric amount of O_2 needed for a complete combustion

(b) The stoichiometric amount of air needed for a complete combustion

(c) The stoichiometric mixture by volume (% ethane in the stoichiometric mixture)

(d) The A/F ratio (in a mass/mass ratio).

6. Soil venting, coupled with granular activated carbon (GAC) for off-gas treatment, is being considered for a site impacted by PCE (C_2Cl_4). The PCE removal efficiency by GAC is 95% and the discharge limit to atmosphere is 0.05 kg PCE/day. There are two GAC units (2,500 kg GAC in each unit) operated in series and the actual adsorption capacity is 0.2 kg PCE/kg GAC.

(a) Determine the maximum allowable PCE loading rate to the GAC system (in kg/day)

(b) Estimate the breakthrough time if the system is running continuously (in days)

7. Soil venting with direct incineration for off-gas treatment is considered as an alternative to bioremediation at a site. The off-gas (100 scfm) is to be burned at 2,000 °F (1,093 °C) with an additional 20 scfm of supplementary fuel and dilution air. To achieve the 99% destruction, the required residence is one second. The standard condition is T = 77 °F (25 °C) and P = 1 atm. What is the minimum required volume of the incineration chamber?

8. Off-gas from a process is to be treated by GAC adsorbers. The chlorobenzene concentration in the off-gas is 1,500 ppm. The gas flow rate is 20 m^3/min @T = 20 °C. Two 2,000-kg GAC adsorbers are to be used. Determine the maximum amount of chlorobenzene that can be held by one GAC adsorber before exhausted. The adsorption isotherm (T = 20 °C) for chlorobenzene concentrations ranging from 0.0001 to 0.05 psi is (q in kg/kg and P in psi):

$$q = 0.508 P^{0.210}$$

9. The flow rate of the off-gas from a project is 100 m^3/min @T = 77 °F (25 °C) and P = 1 atm. It is to be treated by an incinerator or a catalytic incinerator before discharge. Determine

(a) The minimum required volume of the incineration chamber, if the design retention time is 0.3 second and the temperature in the incinerator is 900 °C

(b) The minimum required volume of the catalyst, if the design space velocity is 30,000/hr.

10. A packed-column absorber is designed to remove 2-butanone from a gas stream (Q = 50 m³/min). Determine
    (a) the minimum liquid-to-air ratio,
    (b) the design liquid-to-air ratio, and
    (c) the design liquid flow rate. Use the following information in calculations:
    - Henry's constant = 0.0274 atm/M
    - absorption factor (AF) = 1.6
    - T = 298 K and P = 1 atm.

11. For the absorber in Question #10, if the influent 2-butanone concentration is 0.3% and the required removal is 60%. The diameter of the column is 0.50 m and K_Ga = 8.0/s. Determine the packing height required

12. Estimate the heat content of the gas stream in Question #10.

13. The acetone concentration in the influent gas stream is 600,000 ppm. How low the operating temperature of a condenser needs to be to achieve a 50% removal efficiency? The Antonine equation for acetone (T = -12.9 to 55.3 °C) is shown below (EPA, 2012):

$$log_{10}(P \; in \; mm - Hg) = 7.11714 - \left[\frac{1210.595}{229.664 + T \; (in \; °C)} \right]$$

14. For a gas stream containing 500 ppm of total petroleum hydrocarbon (TPH), what would be a good size of biofilters to treat this gas stream (flow rate = 10 standard m³/min).

Chapter 8

Control of Nitrogen Oxides Emissions

Nitrogen dioxide (NO_2) is one of the six criteria pollutants. Most of nitrogen oxides in the ambient air come from combustion. Formation of nitrogen oxides in combustion cannot be avoided, but can be reduced. Nitrogen oxides in flue gases can also be removed from back-end treatment.

This chapter starts with an introduction on the family of nitrogen oxides and their sources and emissions (Section 8.1). Section 8.2 describes types of flames, combustion temperature, and stationary combustion systems. Formation mechanisms of three types of NO_x (i.e., thermal, fuel, and prompt) are discussed in Section 8.3. Section 8.4 describes general approaches to reduce NO_x emissions Reduction of NO_x formation can be achieved through combustion modifications (Section 8.5) and fuel switching (Section 8.7). Section 8.6 presents two back-end treatment technologies (i.e., selective non-catalytic reduction and selective catalytic reduction) to reduce NO_x in the flue gas to nitrogen molecules before discharge. Process selection and design calculations with regards to control of nitrogen oxides emissions are presented in Section 8.8.

8.1 Introduction

8.1.1 Nitrogen oxides and NO_x
Nitrogen oxides represent a family of seven compounds: nitrous oxide (N_2O), nitric oxide (NO), dinitrogen dioxide (N_2O_2), dinitrogen trioxide (N_2O_3), nitrogen dioxide (NO_2), dinitrogen tetroxide (N_2O_4), and dinitrogen pentoxide (N_2O_5). EPA regulates NO_2 as a surrogate for this family of nitrogen oxides because it is the most prevalent form in the ambient air. Recently, emissions of N_2O started being separately regulated, because it is a potent greenhouse gas (GHG).

In atmospheric chemistry, NO_x is a generic term for nitrogen oxides that are most relevant to air pollution. In EPA (1999), NO_x includes all seven nitrogen oxides; while nitrous oxide is specifically excluded from NO_x in EPA (2014). Most people and literature consider NO_x is an acronym for the sum of nitric oxide and nitrogen dioxide; that is "$NO_x = NO + NO_2$".
Nitric oxide (NO) is the main form of NO_x emitted from combustion sources. Once emitted to the atmosphere, NO will be rapidly oxidized to NO_2.

Dinitrogen tetroxide (N_2O_4) is in a constant equilibrium with NO_2, and it is considered as part of NO_2 and not regulated separately. While properties of NO_2 and N_2O_4 are very similar, those of NO are quite different. NO_2 is water soluble and has a reddish brown color, while NO is relatively insoluble and colorless. In addition, NO does not absorb UV light so that it cannot initiate photochemical reactions. Nitrous oxide (N_2O), also called laughing gas, dose not absorb UV light and does not participate in photochemical reactions for ozone formation. However, it is one of the major GHGs and also has a significant impact on the ozone layer (EPA, 2014).

Similar to hydrogen sulfide (H_2S) which is in a reduced state when compared to SO_2, ammonia (NH_3) is in a reduced state (valence state of nitrogen = -3) when compared to NO_2 (valence state of nitrogen = 4). Under ambient conditions, pure ammonia is a colorless, pungent-smelling, and caustic gas. It is highly soluble in water to form ammonium hydroxide (NH_4OH) as:

$$NH_{3(g)} + H_2O_{(aq)} \rightarrow NH_4OH_{(aq)} \rightarrow NH_{4(aq)}^+ + OH_{(aq)}^- \qquad (8.1)$$

Ammonia is commonly used as a reducing agent to reduce NO_x in the flue gas to nitrogen molecules before discharge.

8.1.2 Sources and emissions of NOx

Nitric oxide can be produced during thunderstorms from lightening that split nitrogen molecules. Other natural sources, such as fires and vegetation, can also emit nitrogen oxides into atmosphere. Most NO_x in the atmosphere originate from anthropogenic sources; and almost exclusively from combustion sources and nearly all combustion sources emit NO_x.

According to EPA's National Emissions Inventory (NEI), the total amount of NO_x emissions of the U.S. is about 10.8 million tons in 2017. Combustion accounts for the majority, >85%, of the total emission; in which more than 2/3 is from transportation and the remaining is from stationary fuel combustion. Industrial and other processes accounts for 11.9%, while wildfires and other miscellaneous sources contributes 2.8% of the total (Table 8.1). For the stationary combustion sources, boilers in the electric power plants produce about 40%, and industrial boilers, incinerators, gas turbines, engines in various industries contribute the other 60% (EPA, 2014)

Table 8.1 - Sources and amounts of the U.S. NO$_x$ emissions in 2017

Source Category	(1,000 tons)	(%)
Stationary fuel combustion	**2,839**	**26.3**
Electric utility	1,155	10.7
Industral	1,143	10.6
Others	541	5.0
Industrial and other Processes	**1,282**	**11.9**
Chemical & allied product manufacturing	47	0.4
Metals processing	70	0.6
Petroleum & related industries	717	6.7
Other industrial processes	330	3.1
Solvent utilization	1	0.0
Storage & transport	6	0.1
Waste disposal & recycling	110	1.0
Transportation	**6,355**	**59.0**
Highway vehicles	3,695	34.3
Off-Highway	2,660	24.7
Miscellaneous	**301**	**2.8**
Wildfires	119	1.1
Others	181	1.7
Total	**10,776**	**100.0**

Figure 8.1 depicts trends of annual NO$_x$ emissions in the U.S. since 1970. As shown, the emissions have been decreasing, especially from 1990. Most of the reductions came from the transportation and the stationary fuel combustion sectors.

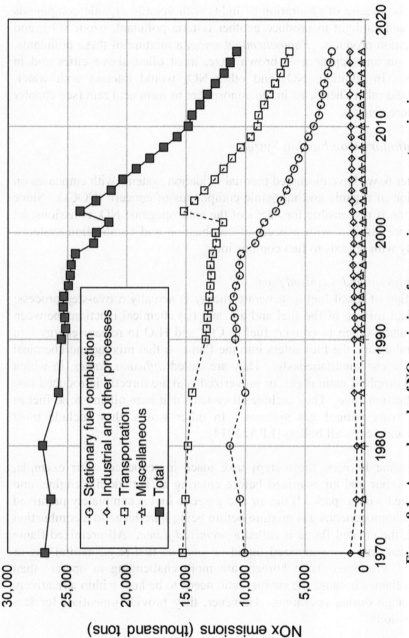

Figure 8.1 - Annual rends of NOₓ emissions from major source categories in the U.S. (1990-2017)

8.1.3 The problems associated with NO_x

NO_2 participates in *photochemical reactions* (i.e., chemical reactions that are initiated as a result of absorption of light) with volatile organic compounds (VOCs) and sunlight to produce another criteria pollutant, ozone (O_3), and other reaction products. *Photochemical smog,* a mixture of these pollutants, would form and appear as a brown haze, most often above cities and in summers. In addition, NO_2 and other NO_x would interact with water, oxygen, and other chemicals in the atmosphere to form acid rain (see chapter 12 for more details).

8.2 Stationary Combustion Systems

In Chapter 6, we have discussed thermal oxidation systems with emphases on destruction of organic and inorganic compounds of concern (COCs). Since combustion is responsible for >85% of the anthropogenic NO_x emissions, let us be more familiar with aspects of combustion and combustion systems, especially with regards to fuel combustion.

8.2.1 Diffusion and premixed flame
Combustion of fossil fuel to generate energy is actually a two-step process: (i) physical mixing of the fuel and air, and (ii) chemical reactions between the fuel and oxygen to convert fuel to CO_2 and H_2O to release energy. In most combustors, the fuel enters into the flame so that mixing and chemical reactions occur simultaneously. They are called *diffusion flames*, in which liquid oil droplets, natural gas, or pulverized coal are directly introduced into the combustion zone. They include all systems that burn oil and solid fuel as well as many natural gas systems. In other words, they include most furnaces and nearly all boilers (EPA, 2014).

But for some burners, these steps take place in sequence. For example, gasoline vapor and air is mixed before entering an automobile engine, and then ignited with a spark. If the air and gaseous fuel is completely premixed to form a homogeneous gas mixture before being introduced to a combustion chamber, the created flame is called a *premixed flame*. All premixed flame burners can only use vaporized liquid or gaseous fuel (e.g., natural gas or propane). Premixed flame burners are more challenging to operate than diffusion flames because the air/fuel ratio needs to be kept within a relatively narrow range during operation. However, they provide potentials for less NO_x emissions.

8.2.2 Combustion temperature

Excess air is the "excess" amount of air, beyond the stoichiometric (theoretical) amount of air needed for complete thermal oxidation of combustible compounds. Figure 8.2 illustrates the combustion temperatures versus the air-to-fuel equivalence ratio. Theoretically, the combustion temperature will be at the maximum when the equivalence ratio is unity (i.e., no excess air). When the equivalence ratio is less than one (i.e., fuel-rich burning), the combustion cannot complete, thus the combustion temperatures are lower. On the other hand, under the fuel-lean conditions, the combustion may complete, but the energy generated from combustion needs to heat up the excess air so that the combustion temperature will also be lower. The combustion temperature and the extent of excess air will have significant impacts on formation of NO_x (and other COCs); and it will be discussed in more details later.

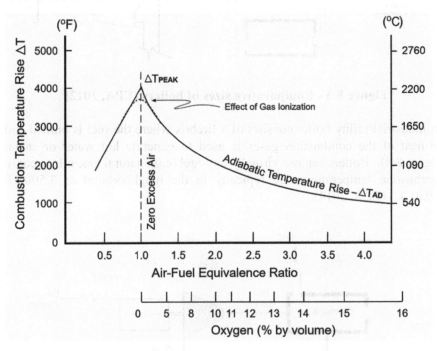

Figure 8.2 - Combustion temperature vs. excess air (modified from EPA, 2012)

8.2.3 Stationary combustion systems

From the perspective of NO_x control, the regulated stationary combustion systems can be classified into three groups: (i) furnaces and boilers, (ii) reciprocating engines, and (iii) combustion turbines. Boilers have a wide range in size from home furnaces to huge industrial/utility boilers (Figure 8.3).

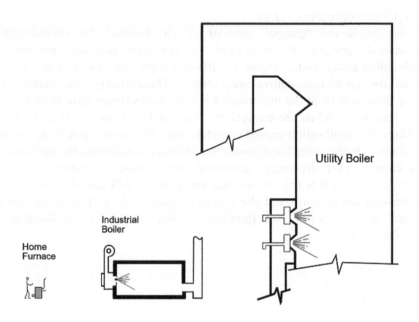

Figure 8.3 - Comparative sizes of boilers (EPA, 2012)

An industrial/utility boiler consists of a firebox where the fuel is burned and the heat of the combustion gases is used to generate hot water or steam (Figure 8.4). Boilers can use virtually any fuel (e.g., natural gas, oil, or coal). Combustion temperatures are typically in the neighborhood of 3,500 °F (1,930 °C) (EPA, 2014).

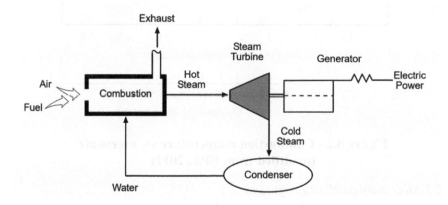

Figure 8.4 - Boiler, turbine, and generator (EPA, 2012)

Reciprocating engines, similar to a car engine, are used for small to medium size stationary power sources, typically < 20 megawatt (MW). Most of them fire natural gas as the primary fuel, and some are capable of using liquid fuel #2 or diesel as alternative. Due to high temperatures, NO_x emissions are higher, roughly ten times the emissions from a boiler burning the same amount of fuel. Reciprocating engines are factory built and their combustion systems and operating controls are not amenable to simple adjust or alternation (EPA, 2014). More details on operation and emission characteristics of reciprocating engines will be covered in Chapter 10 later.

Combustion turbines are basically the aircraft jet engines adapted for stationary power sources. Contrary to reciprocating engines, the combustion in turbines is in a steady-state mode. Air is drawn into a compressor to be pressurized and then enters the combustor to react with fuel to raise the gas stream temperature to the range of about 1,500 to 2,000 °F (815 to 1,090 °C). The higher temperature would increase the volume of the gas stream which is then expanded through a turbine (Figure 8.5). The power generated by the turbine is to run the compressor and for the output shaft power (e.g., for electricity generation). Their power levels can be as large as 250 MW. Although combustion can burn liquid or gaseous fuel, they practically only burn clean fuel to avoid the accumulation of deposits, formed from contaminants in the fuel, on the turbine blades.

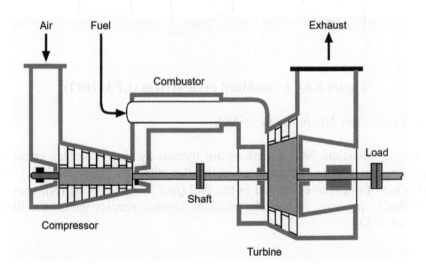

Figure 8.5 - Schematic of a combustion turbine (EPA, 2012)

Thermal efficiency is the fraction of fuel energy content that is converted into shaft power. *A combined cycle* system is to raise the thermal efficiency by using waste heat from the primary engine to drive another power source. As shown in Figure 8.6, the waste heat from a combustion turbine system is used to heat a boiler to drive a steam engine. This arrange could increase the thermal efficiency from ~ 35% of the combustion turbine to ~50% of the combined cycle system (EPA, 2012). When the waste heat from an engine or a turbine is used directly for heating a building or manufacturing process, the system is called *cogeneration*.

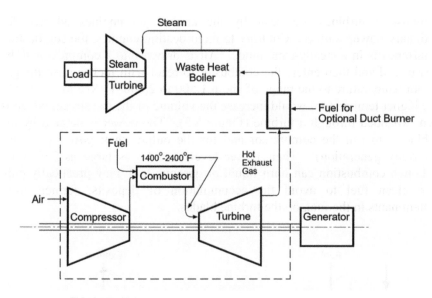

Figure 8.6 - A combined cycle system (EPA, 2012)

8.3 Formation Mechanisms of NO_x

During combustion, NO_x emissions are formed by three complex chemical reactions: (i) thermal fixation of molecular nitrogen (*thermal NOx*), (ii) oxidation of nitrogen contained in the fuel (*fuel NO_x*), and (iii) formation of NO_x due to presence of partially oxidized organic species within the flame (*prompt NO_x*).

8.3.1 Formation of thermal NO_x

When fossil fuel is burned with air in a combustion device, some of the oxygen and nitrogen molecules in the air will combine to form NO due to high temperatures, especially within the burner flame. The simplified overall reaction of thermal NO_x formation is:

$$N_2 + O_2 \rightarrow 2NO \qquad (8.1)$$

Formation of free radicals is important in combustion. A *free radical* is an uncharged molecule having an unpaired valence electron and it is typically highly reactive and short-lived. The free radicals most often involved in combustion reactions are O, H, N, OH, and hydrocarbons that have lost one or more hydrogen atoms (e.g., methyl radical, CH_3) [Note: A dot is often added to the chemical formula of a free radical to represent the unpaired electron, such as •OH for hydroxyl radical and •CH_3 for methyl radical. However, it is also often omitted in literature]. Free radicals are also formed and participate in photochemical reactions in the atmosphere.

During combustion, for example, these free radicals can be formed as:

$$O_2 + M \rightarrow 2O + M \qquad (8.2)$$

$$N_2 + M \rightarrow 2N + M \qquad (8.3)$$

$$H_2O + M \rightarrow H + OH + M \qquad (8.4)$$

An M is included on both sides of Eqs. 8.2 through 8.4. It is to show that another molecule is involved to allow the reaction to occur, but is not chemically changed at the end of the reaction (similar to the role of hydrocarbon plays in ozone formation as a catalyst).

In 1946, Y.B. Zeldovich, a Russian scientist, proposed free-radical chain formation of thermal NO at high combustion temperatures. The mechanism assumes that O radicals attack N_2 to form N radicals (Eq. 8.5). The N radicals can form NO by reacting with O_2 (Eq. 8.6) or hydroxyl radicals (Eq. 8.7):

$$N_2 + O \leftrightarrow NO + N \qquad (8.5)$$

$$N + O_2 \leftrightarrow NO + O \qquad (8.6)$$

$$N + OH \leftrightarrow NO + H \qquad (8.7)$$

267

The reactions shown in Eqs. 8.5 and 8.6 are mainly responsible for formation of NO, while the reaction shown in Eq. 8.7 is only important in near-stoichiometric and fuel-rich combustion zones. The reaction in Eq. 8.6 is considered as the rate-limiting step because of its higher activation energy; and it will become significant only at temperatures greater than ~1,500 °C (2,700 °F) (EPA, 2014).

The main factors affecting the amount of thermal NO_x formed in combustion are the temperature of the flame as well as the residence time and the amount of the combustion gases in the peak temperature zone of the flame. Both NO and NO_2 are present in the effluent gas streams of most combustion resources. NO is the main form of NO_x emitted from traditional combustion devices, and accounts for ≥95% of the total NO_x emissions. However, ≥50% of the NO_x from newer "low-NO_x" sources (e.g., turbines and engines with lean pre-mix combustors) may be NO_2 (EPA, 2009).

Although oxygen concentration and residence time also affect formation of thermal NO_x, temperature is usually the dominating factor. Figure 8.7 illustrates the extent of thermal NO_x formation as a function of temperature. It shows that formation of thermal NO_x peaks between 1,900 and 2,000 °C. It is the temperature range of most stationary combustion sources operate (2009).

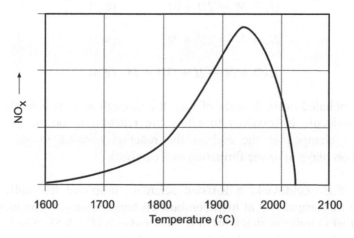

Figure 8.7 - Thermal NO_x formation versus temperature (EPA, 2009)

The vast majority of thermal NO_x is formed in the highest temperature region of the combustion zone. The overall reaction (Eq. 8.1) proceeds rapidly in the forward direction (i.e., formation of NO) at high temperatures. Once the gas begins to cool, the reaction theoretically should reverse. However, it does not happen because of the absence of required free radicals and insufficient activation energy (EPA, 2009).

8.3.2 Formation of fuel NO_x

Fuel NO_x is generated from oxidation of organically-bound nitrogen in fuel during combustion. The amount of fuel NO_x formation depends on the amount of this type of nitrogen in the fuel. U.S. coals contain approximately 0.2 to 3.5 % nitrogen by weight. Anthracite coals constitute the low end, while bituminous coals the high end of this range. When crude oil is distilled, nitrogen tends to accumulate in the heavy residuals; while distillates such as #2 fuel oil and diesel oil typically have little nitrogen. Natural gas contains no organic-nitrogen (EPA, 2009).

Although the fuel NO_x formation would be less at lower flame temperatures, but not affected significantly. The reduction of the oxygen level within the flame, not the average oxygen concentration in the combustion chamber, is critical in reducing fuel NO_x formation (EPA, 2009; EPA 2014).

8.3.3 Formation of prompt NO_x

Prompt NO_x is formed by converting molecular nitrogen into NO via intermediate products with the presence of free radicals such as HCN, NH, and N in the burner flame. These reactions occur in the early phase in the flame front and are not sensitive to the peak gas temperature. While prompt NO_x is part of the total NO_x emissions from practical combustors, it is not normally considered when dealing with control of NO_x emissions (EPA, 2014).

8.4 Approaches to Reduce NO_x Emissions

Techniques used to minimize NO_x emissions from stationary combustion sources can be grouped into three categories: (i) combustion modifications, (ii) add-on (back-end, or end-of-the-pipe) control systems, and (iii) fuel switching. As mentioned, the flame temperature as well as oxygen concentration and residence time affect thermal NO_x formation. The objective of combustion modifications is to create operating conditions that inhibit the NO_x formation reactions or cause the NO_x formed to be reduced back to molecular nitrogen (backward reaction in Eq. 8.1) while the gases are still within the combustor. The add-on NO_x control systems inject chemical reagents to chemically reduce NO_x to molecular nitrogen downstream of the combustors. Fuel switching is to use fuels that would yield lower NO_x emissions because of the specific characteristics of their fuel nitrogen and/or combustion behaviors.

8.5 Combustion Modification

There are five major categories of combustion modification techniques, especially for industrial boilers (EPA, 2003):

1. Low excess-air combustion
2. Off-stoichiometric combustion
3. Flue gas recirculation (FGR)
4. Low NO_x burners
5. Gas reburning

These approaches are to reduce peak temperatures and oxygen concentrations as well as the residence time under these peak conditions so that the formation of thermal NO_x (and fuel NO_x) will minimized.

8.5.1 Low excess-air combustion

For combustion systems, a certain amount of excess air is required to ensure complete combustion of the fuel. Levels of excess air practiced in industrial or utility boilers depend on types of fossil fuel and boiler operating load. The typical excess levels are 3 to 5%, 5 to 20%, and 20 to 50% for gas-fired (near full load), oil-fired, and pulverized coal-fired boilers, respectively (EPA, 2012a). When the excess air is too low, emissions of CO, smoke, and products of incomplete combustion (PICs) in the boiler exhausts will increase (Figure 8.8).

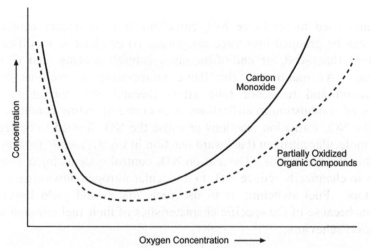

Figure 8.8 - Concentrations of CO and PICs versus oxygen concentrations (EPA, 2003)

There is an optimum excess air level where formation of NO_x is reduced and concentrations of CO and PICs are not excessive (Figure 8.9). To maintain this optimal level, the control system may need to be upgraded (e.g., installation of an oxygen analyzer) so that the air flow can be precisely and continuously controlled (EPA, 2012a). NO_x reductions between 16 and 20% can be achieved on gas- and oil-fired boilers, 20% (on the average) for coal-fired utility boilers (EPA, 2003).

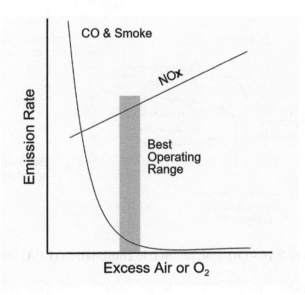

Figure 8.9 - NO_x, CO, and PIC emissions vs. excess air (EPA, 2012)

8.5.2 Off-stoichiometric combustion
The principle of off-stoichiometric combustion is to have two separate zones for combustion of the air and fuel mixtures. In the fuel-rich zone, the fuel is fired with less than the stoichiometric amount of air. The combustion temperature will be lower because the combustion will not be complete. The other zone is air-rich where the balance of the combustion air is introduced to combust the fuel (Figure 8.10). Both thermal and fuel NO_x formation can be reduced because oxygen-deficiency in the fuel-rich zone and temperatures in both the fuel-rich and air-rich zone are lower than that in single-stage combustion. As shown, there are primary air and secondary air added to the combustion system. *Primary air* is the air supplied to the burners with the fuel to control the rate of combustion as well as the amount of fuel that can be burned. *Secondary air* is the air added to control the combustion efficiency and/or completeness of fuel burned.

271

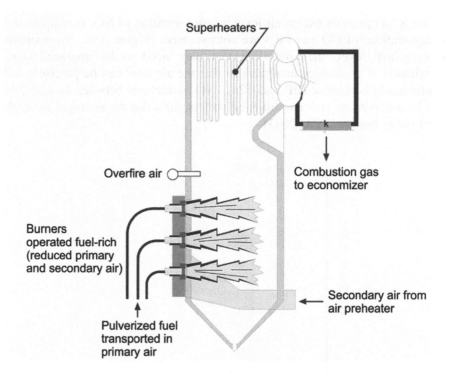

Figure 8.10 - Off-stoichiometric combustion (EPA, 2003)

Off-stoichiometric combustion can be accomplished by employing over-fire air (OFA) ports. They are air injection nozzles located above the burners. The burners are operated fuel-rich while the OFAs, working as air-rich, maintain the remainder of the combustion. Figure 8.11(a) illustrates that in some boilers the burners are operated in a staggered operation, called *biased firing*; with some burners in the fuel-rich mode employing a low stoichiometric ratio (SR) while the others in the air-rich mode. When some burners are operated on air only (the air-rich mode), this modification is called *burners-out-of-service* (BOOS) as shown in Figure 8.11(b).

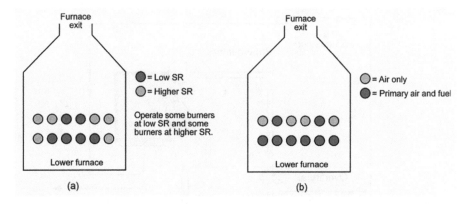

Figure 8.11 - Burner arrangement: (a) Biased firing; (b) Out-of-service (EPA, 2012)

The NO_x emission reductions, using off-stoichiometric combustion, are approximately 30 to 40% for the gas- and oil-fired boilers and 30 to 50% for the coal-fired boilers.

8.5.3 Flue gas recirculation (FGR)

Flue gas recirculation (FGR), also called *exhaust gas recirculation* (EGR) on reciprocating engines, works by recycling portion (10 to 30% typical) of the flue gas back to the main combustion chamber (Figure 8.12). The FGR operates without changing the flow rates of the inlet air, the fuel, and the exhaust. However, the flow rate through the combustor is increased. Increasing the gas flow (and the mass) in the combustion zone decreases the combustion temperature because the same amount of supplied energy is distributed to a larger mass. The lower oxygen in the recycle stream also dilutes the oxygen concentration in the combustion zone. The reductions of the temperature and oxygen concentration lower the NO_x formation.

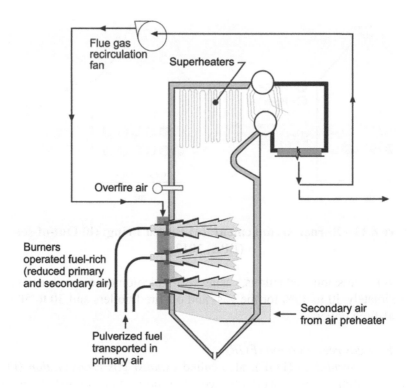

Figure 8.12 - Flue gas recirculation (EPA, 2012)

FGR has been used to reduce thermal NO_x emissions from large gas-, oil-, and coal-fired boilers. NO_x emission reduction of 40 to 50% is achievable with recirculation of 20 to 30% of the exhaust gas in gas- and oil-fired boilers. If the recirculation rate is too high, the flame may become unstable so that emissions of CO and PICs may increase (EPA, 2003). The extents of NO_x emission reduction of the coal-fired boilers employing FGR are generally less.

8.5.4 Low NO_x burner
The approach of using BOOS for off-stoichiometric combustion will not be doable for small boilers with one or two burner; and many smaller boilers may not have a proper location for installation of OFA ports. Many manufacturers have developed new designs on burners themselves with goals of reducing NO_x formation. Some low-NO_x burners are designed to have fuel-rich and air-rich regions (i.e., similar to the approach of off-stoichiometric combustion), while the others are to control the flame shape to minimize the reaction between nitrogen and oxygen at peak flame temperatures (EPA, 2012a).

Schematic of a dual-register low-NO$_x$ burner is illustrated in Figure 8.13, as an example. For this burner, the mixture of pulverized coal and primary air is controlled to delay the fuel combustion slightly (i.e., to create fuel-rich). The remaining of the combustion air is introduced as the secondary air through two concentric air zones that surround the nozzle (air-rich). The flame produced by this low-NO$_x$ nozzle is elongated, compared to the flame produced by a conventional burner. Formations of both thermal and fuel NO$_x$ are reduced due to the lower peak flame temperature and less availability of oxygen in the flame (EPA, 2003). Most low-NO$_x$ burners have demonstrated NO$_x$ reduction of 25 to 40%, while some reportedly > 60% (EPA, 2004)

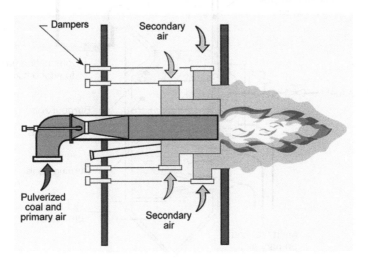

Figure 8.13 - A dual-register low NO$_x$ burner (EPA, 2003)

8.5.5 Gas reburning

Figure 8-14 illustrates a gas reburning system. In addition to the coal-fired burners located at the bottom (the "burning" zone), separate gas-fired burners are added in the upper portion of the combustion chamber (the "reburn" zone). In the return zone, more fuel (i.e., natural gas) is injected without accompanying air to create a fuel-rich zone. Within this zone, no additional NO$_x$ will be formed due to lack of oxygen and lower temperatures. The temperatures are below the threshold for NO$_x$ formation, but sufficient to ignite the supplemental fuel. The hungry fuel species will then take oxygen from previously-formed NO$_x$ by reducing it to nitrogen molecules. OFA can then be added to the "burnout" zone to combust the PICs generated in the reburn zone. Not much NO$_x$ will be formed in this zone because temperatures will not be high (the heat is being lost to the wall of the furnace). Gas reburning is essentially a practice of off-stoichiometric combustion (EPA, 2014).

All types of existing coal- and oil-fired units can employ gas reburning, provided they have enough room above their original burners for installation of gas burners and the OVA ports. 50 to 70% of NO_x reductions at the full boiler load have been demonstrated (EPA, 2003).

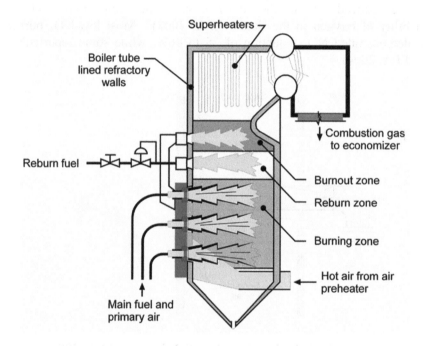

Figure 8.14 - A gas reburning system (EPA, 2003)

8.6 Flue-gas Treatment

In addition to combustion modification to reduce NO_x formation, the formed NO_x can be removed from the exhaust (flue-gas) stream by back-end control. Two techniques that are commonly-used in stationary boilers to reduce NO_x in the flue gas to nitrogen molecules before discharge to atmosphere. They are *selective non-catalytic reduction* (SNCR) and *selective catalytic reduction* (SCR) processes. In addition to these two commercially-available approaches, there are some emerging technologies employing principles of reduction, absorption, or oxidation (EPA, 1999; EPA, 2014).

8.6.1 Selective non-catalytic reduction (SNCR)

There are several commercial versions of SNCR. In general, a more reduced nitrogen compound (mostly, ammonia (NH_3) or urea (CH_4N_2O)) is injected into the post combustion zone of a boiler (Figure 8.15). The injected ammonia or urea will react with NO and NO_2 in a complex set of high-temperature gaseous phase reactions to reduce NO_x to nitrogen molecules. The process is called "selective reduction" because of the use of a reagent to react with NO_x. The overall reactions for NO reduction are:

$$NO + NH_3 + \frac{1}{4}O_2 \rightarrow N_2 + \frac{3}{2}H_2O \quad (8.8)$$

$$NO + \frac{1}{2}CH_4N_2O + \frac{1}{4}O_2 \rightarrow N_2 + H_2O + \frac{1}{2}CO_2 \quad (8.9)$$

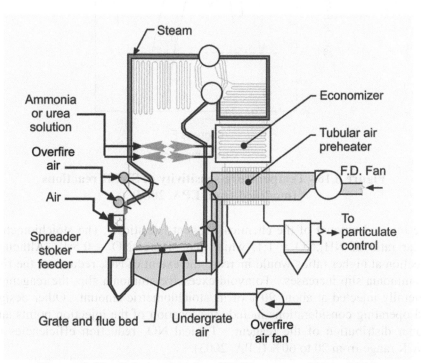

Figure 8.15 - Schematic of an SNCR system (EPA, 2003)

The reactions shown in Equations 8-8 and 8-9 are very temperature sensitive. As shown in Figure 8.16, the best conversion occurs in a temperature range of approximately 1,600 to 1,900 °F (870 to 1,870 °C). Above this temperature range, ammonia or urea will be increasingly oxidized to form NO_x. Below this temperature range, more ammonia will pass through the system unreacted to generate ammonia emissions called ammonia slip. An ammonia slip with an ammonia concentration >10 ppm is undesirable because ammonia can react with chlorine or sulfur compounds to form a light-scattering particulate plume.

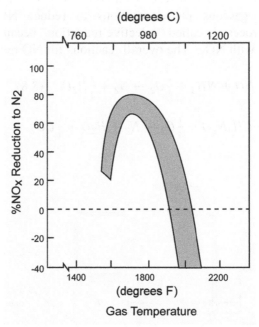

Figure 8.16 - Temperature-sensitivity of SNCR reactions (modified from EPA, 2003)

The injected amount of the chemical reagent is critical. The stoichiometric molar ratio of $NH_3:NO$ = 1:1, while that of urea:NO = 0.5:1. Although injection at higher ratios would increase the extent of NO_x reduction, the risk of ammonia slip increases. To avoid excessive ammonia slip, the reagent is generally injected at about 90% of its stoichiometric amount. Other design and operating considerations include the location of the injection points and proper distribution of the reagent. Typical NO_x reduction efficiencies of SNCR range from 20 to 60% (EPA, 2003).

8.6.2 Selective catalytic reduction (SCR)
Similar to SCNR in principle, a catalyst module is added in SCR (Figure 8.17). The main function of the catalyst is to reduce the temperature required for chemical reduction reactions (Eqs. 8.8 & 8.9). Typical catalysts include vanadium pentoxide (V_2O_5), titanium dioxide (TiO_2), tungsten trioxide (WO_3) and noble metals. The temperature needed for efficient reduction

278

would depend on the type of catalyst used, typically in the range of 550 to 750 °F (290 to 400 °C). Temperature variation of as little as ± 50 °F (±30 °C) can have an impact on its NO_x reduction performance. SCR systems could deliver NO_x reduction in the range of 60 to 90% (EPA, 2003). The ammonia oxidation at higher temperatures and ammonia slips at lower temperatures, as those in SNCR, are also of concern.

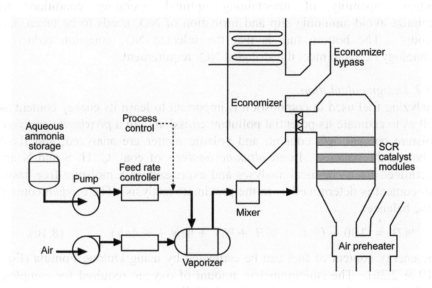

Figure 8.17 - Schematic of an SCR system (EPA, 2003)

8.7 Fuel Switching

Fuel NO_x formation in boilers can be reduced by fuel switching. Three potential approaches are (1) conversion of fuel supply with a lower nitrogen content or the one with fuel nitrogen in a chemical form less likely to form NO_x; (2) firing the boiler with natural gas (which has no organically-bound nitrogen) during the periods when the ambient ozone levels are high (e.g., hot summer days), and (3) simultaneous combustion of coal and gas.

8.8 Process Selection and Design Considerations

8.8.1 Process selection

As discussed, there are three different NO_x formation mechanisms: fuel, thermal, and prompt. The NO_x emission reductions can be achieved mainly by fuel switching, flue-gas treatment, and combustion modifications. They are not exclusive to others. A combined approach can be applied, if it is found effective. In addition to economic consideration, all these approaches have their advantages and limitations. Fuel switching is mainly to reduce fuel NO_x, not thermal and prompt NO_x. The cost, availability of fuel with low nitrogen content, and effectiveness in NO_x emission reduction should be the main considerations. For combustion modifications, feasibility of upgrading the existing combustor, the cost of the upgrade or purchasing new devices, and their effectiveness are main considerations. For the back-end control, capability of maintaining optimal operating conditions to minimize/avoid ammonia slip and formation of NO_x needs to be taken into account. The bottom line is that the selected NO_x emission reduction technology needs to meet the stringent NO_x requirement.

8.8.2 Design calculation

Analyzing fuel used in combustion is important to learn its energy content as well as to estimate its potential pollutant emissions. In a *proximate analysis*, moisture content, ash content, and volatile matter are analyzed; and fixed carbon is the balance. In an *ultimate analysis* of coal, C, H, S and N are determined from chemical analyses and expressed on a moisture-free basis. Ash content is determined as in the proximate analysis. The oxygen content is the balance:

$$\%O = 100 - (\%C + \%H + \%S + \%N + \%Ash) \qquad (8.10)$$

The energy content of fuel can be estimated by using Dulong'Fomula (Eqs. 2.19 & 2.20). The stoichiometric amount of oxygen required for complete combustion can be estimated by using the following equations:

$$C + O_2 \rightarrow CO_2 \qquad (8.11)$$

$$H + \frac{1}{4}O_2 \rightarrow \frac{1}{2}H_2O \quad (8.12)$$

$$S + O_2 \rightarrow SO_2 \qquad (8.13)$$

$$N + O_2 \rightarrow NO_2 \qquad (8.14)$$

As an example, Eq. 8.11 indicates that one mole of O_2 will be needed to combust one mole of C; and the reaction will generate one mole of CO_2. Therefore, the stoichiometric oxygen amount for complete combustion will be (on a molar basis):

$$moles\ of\ O_2 = moles\ of\ \left\{ C + \frac{1}{4}H + S + N - \frac{fuel\ O}{2} \right\} \qquad (8.15)$$

Assuming oxygen concentration in the air is 21%, the stoichiometric air amount required for complete combustion can be found as (on a molar basis):

$$moles\ of\ air = \frac{moles\ of\ oxygen}{0.21} \qquad (8.16)$$

Example 8.1 Stoichiometric air for coal combustion

For a coal firing rate of 100 metric ton/hr, what is the minimum amount of air required (in standard m³/hr) to complete the combustion? Results of coal ultimate analysis (by wt.) are shown below:

- C: 65.0%
- H: 8.0%
- O: 7.0%
- S: 1.0%
- N: 1.5%
- Moisture: 4.0%
- Ash content: 13.5%

Solution:

(a) The mass loading rates of C, H, O, S, N, H_2O, and ash are 65, 8, 7, 1, 1.5, 4, and 13.3 ton/hr, respectively.

(b) The molar loading rates of C, H, O, S, N, and H_2O are 5,420, 8,000, 438, 31, 107, and 222 kg-mole/hr, respectively (by dividing the mass loading rate with the corresponding atomic mass or MW).

(c) Use Eq. 8.15 to find the stoichiometric oxygen requirement:

$$O_2 = \left\{5,420 + \frac{8000}{4} + 31 + 107 - \frac{438}{2}\right\} = 7,370 \; kg - mole/hr$$

(d) Use Eq. 8.16 to find the stoichiometric air requirement:

$$stoichiometric \; air = \frac{7,370}{0.21} = 35,100 \; \frac{kg - mole}{hr}$$
$$= (35,100)(24.46) = 8.6 \times 10^5 \; m^3/hr$$

Discussion:

1. In this example, standard temperature = 25 °C (77 °F), typically for combustion. The molar volume of ideal gas = 24.46 m³/kg-mole.

2. If 40% of excess air is used, the air flow rate will be 140% of the value derived in part (d).

Example 8.2 Flue gas flow rate for coal combustion

For the combustion described in Example 8.1, what would be the flow rate of the flue gas (in kg-mole/hr and standard m³/hr)?

Solution:

(a) For each mole of C, H, S, N in the fuel will produce 1, ½, 1, and 1 mole of CO_2, H_2O, SO_2, and NO_2, respectively.

(b) The moles of nitrogen in the influent gas = (79%) of the influent air =

= (0.79)(35,100) = 27,730 kg-mole/hr

(c) Total moles of the flue gas = moles of $\{CO_2, H_2O, SO_2, NO_2, N_2, and O_2\}$

= 5,420 + (8,000/2 + 222) + 31 + 107 + 27,730 + 0 = <u>37,510 kg-mole/hr</u>

(d) Flue gas flow rate = (37,520(24.42) = <u>9.16×10^5 standard m³/hr.</u>

Discussion:

1. The moisture in the flue gas came from two sources: moisture in the coal and water produced from combusting H in the coal.

2. Nitrogen in the supply air is assumed to be inert (i.e., no formation of NO_x).

3. It is assumed that the combustion is complete. There's no oxygen in the flue gas because the oxygen in the influent is at the stoichiometric amount.

4. If excess air is added, similar calculations can be done, but oxygen will be present in the flue gas and the nitrogen will also be in a larger amount.

5. Using a spread sheet is a good idea to solve questions of this nature,

---

Example 8.3 Pollutant concentrations in flue gas of coal combustion

For the combustion described in Example 8.2, what would be concentrations of CO_2 and NO_2 in the flue gas? What would be the mass emission rate of CO_2 (in ton/hr)?

Solution:

(a) From Example 8.3, the molar flow rate of CO_2, NO_2, and flue gas are 5,420, 107, and 37,510 kg-mole/hr, respectively.

$[CO_2]$ = (5,420) ÷ (37,510) = <u>14.45%</u>

$[NO_2]$ = (107) ÷ (37,510) = 2.85×10^{-3} = <u>2,850 ppm</u>

(b) Mass emission rate of CO_2

$= (5,420 \text{ kg-mole/hr})(44 \text{ kg/kg-mole}) = 238,500 \text{ kg/hr} = \underline{238.5 \text{ ton/hr}}$.

Discussion:

1. From these calculations, the CO_2 emission rate of burning 100 ton/hr of coal is 239 ton/hr. It implies that burning one ton of coal will emit ~ 2.4 ton of CO_2.

2. The calculated NO_2 concentration in the flue gas is 2,850 ppm, relatively high. The only NO_2 source is from combusting fuel nitrogen.

For selective non-catalytic reduction of NO, the stoichiometric molar ratio of NH_3: NO = 1:1. With a design molar ratio and the mass loading rate of NO, the rate of ammonia addition can be readily estimated. To avoid excessive ammonia slip, ammonia is generally injected at about 90% of its stoichiometric amount. Other design and operating considerations include the location of the injection points and proper distribution of the reagent. Typical NO_x reduction efficiencies of SNCR range from 20 to 60% (EPA, 2003).

Example 8.4 Ammonia addition of an SNCR system

A boiler employs an SNCR system to reduce its NO_x emissions. The NO_x concentration at the boiler outlet is 200 ppm. A pilot study was conducted and found that dosing ammonia at 90% of the stoichiometric NH_3: NO ratio would reduce 40% of NO_x emissions. Calculate the ammonia addition rate if the flue gas flow rate is 3,000 standard m³/min.

Solution:

(a) MW of NO = 14 + 16 = 30. At 25 °C,

$$200 \, ppm = (200) \left[\frac{30}{24.46} \right] = 246 \, mg/m^3$$

(b) The mass loading rate of NO

$= (246 \text{ mg/m}^3)(3,000 \text{ m}^3/\text{min}) = 0.737 \text{ kg/min} = 737 \text{ g/min}$

(c) The molar loading rate of NO

$= (737 \text{ g/min}) \div (30 \text{ g/g-mole}) = 24.6 \text{ g-gmole}$

283

(d) At a molar ratio of 0.9, the molar loading rate of ammonia

$= (0.9)(24.6) = 22.1$ g-mole

(e) The mass feeding rate of NH_3

$= (22.1$ g-mole$)(17$ g/g-mole$) = \underline{376 \text{ g/min}}$

8.9 Summary

1. Three different NO_x formation mechanisms from fuel combustion: thermal, fuel, and prompt NO_x.
2. Most of NO_x in the flue gas is NO; but NO will be oxidized to NO_2 readily after being emitted.
3. Formation of thermal and prompt NO_x can be reduced, but is not totally avoidable. Reduction in formation can be achieved by reducing the combustion temperature and the oxygen concentration in the flame.
4. Combustion modification to reduce thermal NO_x formation can be done with low excess air combustion, off-stoichiometric combustion, flue gas recirculation, low-NO_x burners, and gas reburning.
5. Emission reduction of fuel NO_x can be achieved by fuel switching.
6. Back-end control can reduce NO_x to molecular nitrogen. Two common approaches are selective non-catalytic reduction and selective catalytic reduction. Ammonia and urea are commonly-used as the reducing agents.

Bibliography

de Nevers, N (2000). *Air Pollution Control Engineering (2nd edition)*, McGraw-Hill Companies, Inc.

Kuo, J. (2015). Air Quality Issues Related to Using Biogas from Anaerobic Digestion of Food Waste, California Energy Commission, CEC-500-2015-037, March 2015, 73 pages.

Kuo, J.; Dow, J. (2017). *Biogas Production from Anaerobic Digestion of Food Waste and Relevant Air Quality Implication, J. Air & Waste Management Association*, 67(9), 1000-1011. doi: 10.1080/10962247.2017.1316326.

USEPA (1999). *APTI 444: Air Pollution Field Enforcement - Student Manual*, Office of Air Quality Planning and Standards, United States Environmental Protection Agency, Research Triangle Park, NC 27711.

USEPA (1999). *Nitrogen Oxides (NO$_x$), Why and How They Are Controlled*, EPA 456/F-99-006R, Office of Air Quality, United States Environmental Protection Agency, Research Triangle Park, NC 27711.

USEPA (2002). *EPA Air Pollution Control Cost Manual (6th edition)*, EPA/452/B-02-001, Office of Air Quality Planning and Standards, United States Environmental Protection Agency, Research Triangle Park, NC 27711.

USEPA (2003). *APTI 452: Principles and Practices of Air Pollution Control - Student Manual (3rd edition)*, Air Pollution Training Institute, United States Environmental Protection Agency, Research Triangle Park, NC 27711.

USEPA (2003). *APTI 455: Inspection of Gas Control Devices and Selected Industries - Student Manual (2nd revision)*, Air Pollution Training Institute (APTI), Environmental Research Center, United States Environmental Protection Agency, Research Triangle Park, NC 27711.

USEPA (2003b). *Selective Non-Catalytic Reduction (SNCR)*, Air Pollution Control Technology Fact Sheet, EPA-452/F-03-031, United States Environmental Protection Agency.

USEPA (2003c). *Selective Catalytic Reduction (SCR)*, Air Pollution Control Technology Fact Sheet, EPA-452/F-03-032, United States Environmental Protection Agency.

USEPA (2009). *APTI 418: NO$_x$ Emission Control from Stationary Sources - Student Manual*, prepared by NESCAUM for United States Environmental Protection Agency.

USEPA (2012). *APTI 415: Control of Gaseous Emissions - Student Guide*, Office of Air and Radiation, United States Environmental Protection Agency, Research Triangle Park, NC 27711.

USEPA (2012). *APTI 427: Combustion Source Evaluation - Student Manual (3rd Edition)*, prepared by ICES Ltd. for Air Pollution Training Institute, United States Environmental Protection Agency, Research Triangle Park, NC 27711.

USEPA (2014). *APTI 418: NO$_x$ Emission Control from Stationary Sources - Student Manual*, prepared by NESCAUM in 2009 (revised by Doyle B.W. & Slot S. in 2014) for United States Environmental Protection Agency.

Exercise Questions

1. For a coal firing rate of 120 metric ton/hr, what is the minimum amount of air required (in standard m^3/hr) to complete the combustion? Results of coal ultimate analysis (by wt.) are shown below:
   - C: 60.0%
   - H: 8.0%
   - O: 9.0%
   - S: 2.0%
   - N: 1.0%
   - Moisture: 5.0%
   - Ash content: 16.0%

2. For the coal combustion described in Question #1, what would be the flow rate of the flue gas (in kg-mole/hr and stand m^3/hr)?

3. For the coal combustion described in Question #1, what would be the concentrations of CO_2 and NO_2 in the flue gas? What would be the mass emission rate of CO_2 (in ton/hr)?

4. For the coal combustion described in Question #1, if 50% excess air is applied, what would be the flow rate of the flue gas (in kg-mole/hr and stand m^3/hr)?

5. For the coal combustion described in Question #1, if 50% excess air is applied, what would be the concentrations of CO_2 and NO_2 in the flue gas? What would be the mass emission rate of CO_2 (in ton/hr).

6. A boiler employs an SNCR system to reduce its NO_x emissions. The NO_x concentration at the boiler outlet is 300 ppm. A pilot study was conducted and found that dosing ammonia at 95% of the stoichiometric NH_3: NO ratio would reduce 45% of NO_x emissions. Calculate the ammonia addition rate if the flue gas flow rate is 2,500 standard m^3/min.

Chapter 9

Control of Sulfur Oxides Emissions

Sulfur dioxide (SO_2) is one of the six criteria pollutants. More than 70% of sulfur dioxide in the atmosphere is from fossil fuel combustion, especially from the stationary sources and 20% from industrial processes. Eliminating or minimizing content will greatly reduce the emission of sulfur dioxide. The sulfur oxides in the flue as can also be removed by scrubbing.

This chapter starts with an introduction on the impacts, sources and rates of SO_2 emissions as well an overview of emission reduction technologies (Section 9.1). Section 9.2 presents chemistry of sulfur and sulfur oxides. Flue gas desulfurization (FGD) is the approach to remove sulfur oxides from flue gas. Seven wet FGD technologies are presented in Section 9.3, while Section 9.4 presents two dry FGD approaches. Five alternatives to the end-of-the pipe emission reduction technologies are described in Section 9.5. Process selection and design calculations are then covered in Section 9.6.

9.1 Introduction

9.1.1 The problems
Sulfur dioxide (SO_2) is used as the indicator of a larger group of gaseous sulfur oxides (SO_x) because its much higher concentrations in the ambient air. Most people consider SO_x is the sum of SO_2 and sulfur trioxide (SO_3), but some include sulfuric acid (H_2SO_4) in this group.

SO_x can exert adverse impacts to human health and the environment. Short-exposure to SO_x can hurt human respiratory system and make breathing difficult, especially to sensitive receptors. At high concentrations, they can harm vegetation by damaging foliage and decreasing growth. They can also react with other compounds in the atmosphere to form fine particles that contribute to particulate matter (PM) pollution causing additional health problem, damaging façade of buildings, statues, and monuments, and impairing visibility. They are also major contributors to acid rain which can harm sensitive ecosystem.

9.1.2 Sources of the problems

All the fossil fuels contain sulfur compounds to some extent. Table 9.1 tabulates the sources and emissions of SO_2 in the U.S. in 2017. The total SO_2 emission rate is about 2.8 million tons in 2017. Burning of fossil fuel by power-plant boilers and industrial boilers represents more than two-thirds of anthropogenic SO_2 emissions into the atmosphere. The largest non-combustion sources are industrial processes (19.0%), such as extracting metal from ore (e.g., copper smelting), petroleum refinery operations, and chemical and allied product manufacturing. Transportation accounts for 3.4% of the total emissions, while miscellaneous sector, including wildfires, accounts for the balance, 5.3%.

Table 9.1 - SO_2 emission rates and source categories in the U.S. in 2017

Source Category	(1,000 tons)	(%)
Stationary fuel combustion	**2,035**	**72.3**
Electric utility	1,385	49.2
Industral	534	19.0
Others	116	4.1
Industrial and other Processes	**534**	**19.0**
Chemical & allied product manufacturing	123	4.4
Metals processing	105	3.7
Petroleum & related industries	104	3.7
Other industrial processes	167	5.9
Solvent utilization	0	0.0
Storage & transport	3	0.1
Waste disposal & recycling	32	1.1
Transportation	**96**	**3.4**
Highway vehicles	27	1.0
Off-Highway	69	2.4
Miscellaneous	**150**	**5.3**
Wildfires	71	2.5
Others	79	2.8
Total	**2,815**	**100.0**

Figures 9.1 depicts trends of annual SO_2 emissions in the U.S. since 1970. As shown, the emissions have been decreasing, especially from 1990. Most of the reductions came from the stationary fuel combustion and the industrial sectors.

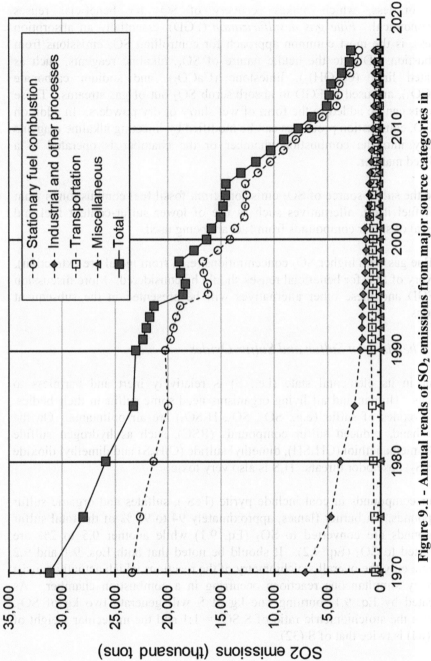

Figure 9.1 - Annual rends of SO$_2$ emissions from major source categories in the U.S. (1990–2017)

9.1.3 Overview of control alternatives to the problems

Typical SO_x concentrations in flue gases from coal- or oil-burning boilers are 0.1% or less, which makes recovery of SO_2 for beneficial reuses un-economical. *Flue gas desulfurization* (FGD), essentially an absorption process, is the most common approach for controlling SO_x emissions from combustion. Due to the acidic nature of SO_x, alkaline reagents, such as hydrated lime $(Ca(OH)_2)$, limestone $(CaCO_3)$, and sodium carbonate (Na_2CO_3), are used in FGD to absorb/scrub SO_2 out of gas streams. These reagents can be added in the form of wet slurry or dry powders. In addition to FGD, combustion processes can be modified by injecting alkaline reagents directly into the combustion chamber or the chamber is operated in a fluidized manner.

Since the sulfur source of SO_2 emissions from fossil fuel combustion is from fossil fuel itself, alternatives such as use of lower sulfur-content fuel and removal of sulfur compounds from fuel are being used.

For flue gas with higher SO_2 concentrations (e.g., from metal ore extraction), recovery of SO_2 for beneficial reuses should be considered. More discussion on FGD and these other alternatives will be presented in the subsequent sections.

9.2 Chemistry of Sulfur and Sulfur Oxides

Sulfur in its elemental state (i.e., S) is relatively inert and harmless to humans. Human and all living organisms need some sulfur in their bodies. Many oxides of sulfur (e.g., SO_2, SO_3, H_2SO_4) are air pollutants. On the other hand, reduced sulfur compounds (RSC), such as hydrogen sulfide (H_2S), methanethiol (CH_3SH), dimethyl sulfide (C_2H_6S) and dimethyl dioxide $(C_2H_6S_2)$ pose odor threats. H_2S is also very toxic.

Sulfur compounds in coal include pyrite (FeS_2), sulfates and organic sulfur compounds. In burner flames, approximately 94 to 95 % of the total sulfur compounds are converted to SO_2 (Eq. 9.1) while another 0.5 to 2% are converted to SO_3 (Eq. 9.2). It should be noted that both Eqs. 9.1 and 9.2 indicate only the overall stoichiometry of the reactions which actually consist of many simultaneous reactions occurring in a combustion chamber. As indicated by Eq. 9.1, burning one kg of S will generate two kg of SO_2, because the stoichiometric ratio of $S:SO_2 = 1:1$ and the molecular weight of SO_2 (64) is twice that of S (32).

Although SO_2 (a polar compound) is more soluble than SO_3 (a non-polar compound) in water under ambient conditions, SO_3 will preferentially combine with a water molecule to form sulfuric acid (H_2SO_4) when the flue gas is cooled below approximately 600 °F (316 °C) (Equation 9.3):

$$S + O_2 \rightarrow SO_2 \quad (9.1)$$

$$S + \frac{1}{2}O_2 \rightarrow SO_3 \quad (9.2)$$

$$SO_3 + H_2O \rightarrow H_2SO_4 \quad (9.3)$$

Figure 9.2 illustrates the mass balance and fate and transfer of fuel sulfur in a boiler. As shown, about 3 to 5.5% of fuel sulfur will not be converted to SO_x and they will leave the boiler with the bottom ash. Changes in the boiler operating conditions (e.g., boiler oxygen levels and burner characteristics) will not significantly affect the fraction of fuel sulfur leaving with the ash, but they will influence the ratio of SO_2 to SO_3 (EPA, 2003).

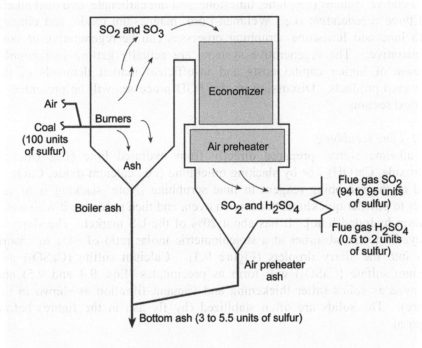

Figure 9.2 - Fate and transfer of fuel sulfur in a boiler (EPA, 2003)

9.3 Wet Flue Gas Desulfurization

Many different FGD processes have been developed. FGD processes can be wet or dry. Wet FGD processes use a liquid absorbent and they can remove >90% of SO_2 emissions. A dry FGD process is basically a two-step process which uses a wet or dry spray of alkaline reagents to react with SO_2 and form dry particles, which are then collected in a fabric filter, ESP, or cyclone. Removal efficiencies of dry FGD systems range from ~50% to >90%, depending on the reagent used and the system configuration/operation. Wet FGD systems are more common which represent approximately 70% of uses for boilers (EPA, 2012).

The FGD systems can be further grouped into non-regenerative and regenerative. The non-regenerative systems produce a sludge/waste that needs to be properly disposed of, and sometimes referred to as throw-away FGD processes. On the other hand, the regenerative FGD processes generate salable end-products such as concentrated SO_2, H_2SO_4, elemental S, and gypsum (EPA, 2012).

This section describes seven common wet FGD processes: four non-regenerative systems (i.e., lime, limestone, sodium carbonate, and dual alkali) and three regenerative (i.e., Wellman-Lord, magnesium oxide, and citrate). Both lime and limestone scrubbing processes can be regenerative or non-regenerative. The regenerative systems are actually getting less popular because of higher capital costs and insufficient market demands of the recovered products. Discussion on dry FGD processes will be presented in the next section.

9.3.1 Lime scrubbing
An alkaline slurry, prepared directly from hydrated lime (i.e., calcium hydroxide, $Ca(OH)_2$) or by slacking quicklime (i.e., calcium oxide, CaO), is used as the scrubbing reagent in lime scrubbing [Note: slacking is to add water to powder quicklime, put it in an oven, and then pulverize it with water to form hydrated lime]. It has about 20% of the US market. The slurry is sprayed into the absorber at a stoichiometric molar ratio of ~1:1 to absorb SO_2 into the slurry droplets (Figure 9.3). Calcium sulfite ($CaSO_3$) and calcium sulfate ($CaSO_4$) will form as precipitates (Eqs. 9.4 and 9.5) and removed as solids (after thickening and vacuum filtration as shown in the figure). The solids are often stabilized (by fly ash in the figure) before disposal.

$$SO_2 + Ca(OH)_2 \rightarrow CaSO_3 + H_2O \qquad (9.4)$$

$$CaSO_3 + \frac{1}{2}O_2 \rightarrow CaSO_4 \qquad (9.5)$$

It should be mentioned that it is usually not desirable to have a mixture with significant quantities of both calcium sulfite and sulfate for final disposal. Therefore, the oxidation of calcium sulfite to calcium sulfate (i.e., Eq. 9.5) is either enhanced or suppressed).

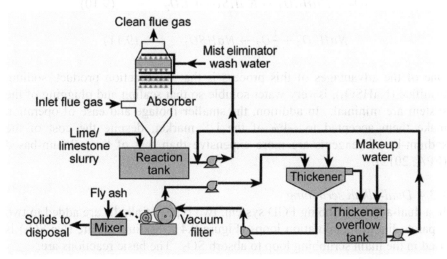

Figure 9.3 - Schematic of a lime-scrubbing FGD process (EPA, 2003)

9.3.2 Limestone scrubbing
The limestone scrubbing FGD process is essentially the same as that of lime scrubbing, except that limestone ($CaCO_3$) is used. The limestone is first crushed and ball-milled to finer than 200 mesh (<0.074 mm), before made into a slurry with water. It represents ~45% of the US market. The basic reactions are:

$$SO_2 + CaCO_3 \rightarrow CaSO_3 + CO_2 \qquad (9.6)$$

$$CaSO_3 + \frac{1}{2}O_2 \rightarrow CaSO_4 \qquad (9.7)$$

9.3.3 Sodium carbonate scrubbing
Similar to both lime and limestone FGD systems, except soluble sodium compounds (i.e., sodium hydroxide (NaOH) or sodium carbonate Na_2CO_3) are used. The reagent is added to a recirculation tank at a rate controlled by the solution pH as well as the content of the dissolved solids. A portion of the recirculation stream is continuously removed, for disposal or treatment, to reduce its fly ash content and the sulfate level. The basic reactions are:

$$SO_2 + 2Na_2CO_3 + H_2O \rightarrow Na_2SO_3 + 2NaHCO_3 \qquad (9.8)$$

$$SO_2 + Na_2SO_3 + H_2O \rightarrow 2\,NaHSO_3 \qquad (9.9)$$

$$SO_2 + NaHCO_3 \rightarrow NaHSO_3 + CO_2 \qquad (9.10)$$

$$NaHSO_3 + \frac{1}{2}O_2 \rightarrow NaHSO_4 \qquad (9.11)$$

One of the advantages of this process is that the reaction product, sodium bisulfate ($NaHSO_3$), is very water soluble so that scaling and plugging of the system are minimal. In addition, the smaller footage and ease of operation make them accepted to ~4% of the US market, despite the cost of the sodium-based reagents are more expensive than that of the calcium-based (EPA, 2012).

9.3.4 Dual-alkali scrubbing
In a dual-alkali scrubbing FGD system, two types of alkalis are added to two separate liquid recirculation loops (Figure 9.4). Sodium sulfite (Na_2SO_3) is used in the main scrubbing loop to absorb SO_2. The basic reactions are:

$$SO_2 + Na_2SO_3 + H_2O \rightarrow 2NaHSO_3 \qquad (9.12)$$

$$NaHSO_3 + \frac{1}{2}O_2 \rightarrow NaHSO_4 \qquad (9.13)$$

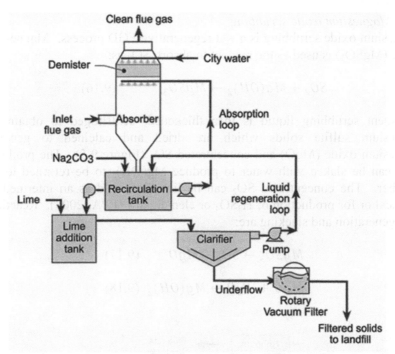

Figure 9.4 - Schematic of a dual alkali-scrubbing FGD process (EPA, 2003)

A portion of the scrubbing fluid is withdrawn to a regeneration loop. In this loop, calcium hydroxide is added to convert the reaction product in Eq. 9.12 (i.e., NaHSO₃) back to sodium sulfite so that it can be returned to the scrubbing loop as the effective reagent. Make-up sodium carbonate (Na₂CO₃) is added to the recirculation loop to compensate for the sodium sulfite that is inadvertently oxidized to sodium sulfate (Eq. 9.13). The byproducts formed in the regeneration loop, calcium sulfite (CaSO₃) and calcium sulfate (CaSO₄), are removed as solids for disposal or beneficial reuses. This type of FGD systems has ~3% of the US market. The basic reactions are:

$$2NaHSO_3 + Ca(OH)_2 \rightarrow Na_2SO_3 + CaSO_3 + 2H_2O \qquad (9.14)$$

$$NaHSO_4 + Ca(OH)_2 \rightarrow NaOH + CaSO_4 + H_2O \qquad (9.15)$$

9.3.5 Magnesium oxide scrubbing

Magnesium oxide scrubbing is a wet regenerative FGD process. Magnesium sulfite ($MgSO_3$) is used as the reagent to absorb SO_2 as:

$$SO_2 + Mg(OH)_2 \rightarrow MgSO_3 \qquad (9.16)$$

The spent scrubbing liquid is then thickened, centrifuged to obtain the magnesium sulfite solids which are dried and calcined to generate magnesium oxide (MgO) and concentrated SO_2 (Figure 9.5). The produced MgO can be slaked with water to produce $Mg(OH)_2$ to be returned to the scrubber. The concentrated SO_2 can be liquefied, sold as an intermediate chemical or for production of H_2SO_4 or elemental S (EPA, 2003). Reactions for regeneration and slacking are:

$$MgSO_3 \rightarrow SO_2 + MgO \qquad (9.17)$$

$$MgO + H_2O \rightarrow Mg(OH)_2 \qquad (9.18)$$

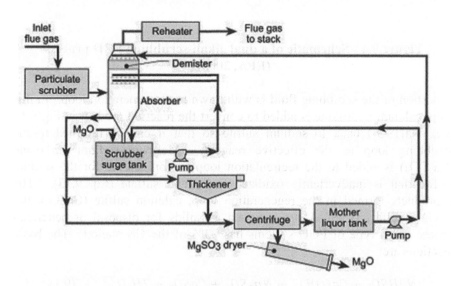

Figure 9.5 - Schematic of an MgO-scrubbing FGD process (EPA, 2003)

9.3.6 Well-Lord scrubbing

The absorption process of Well-Lord scrubbing is identical to the main scrubbing loop of the dual-alkali in which a sodium sulfite solution is used to absorb SO_2 to form sodium bisulfite as:

$$SO_2 + Na_2SO_3 + H_2O \rightarrow 2NaHSO_3 \qquad (9.19)$$

The spent scrubbing solution is sent to an evaporator to convert sodium bisulfite to sodium sulfite (for recycling) and concentrated SO_2 gas (for beneficial reuses) as (EPA, 2003):

$$2NaHSO_3 \rightarrow Na_2SO_3 + H_2O + SO_2 \qquad (9.20)$$

9.3.7 Citrate process

Since SO_2 is acidic, the absorption of SO_2 into water increases as the water pH increases. In this process, a solution of citrate ions ($C_6H_5O_7^{-3}$), citric acid ($C_6H_8O_7$), and sodium thiosulfate ($Na_2S_2O_3$) is used to attain a higher solubility of SO_2 in the solution. The solution is then reacted with H_2S gas to produce elemental S and to regenerate the citrate solution. The H_2S gas can come from petroleum refineries or can be produced on-site by reacting the recovered S with steam and natural gas (EPA, 2012).

9.4 Dry Flue Gas Desulfurization

Dry scrubbing is basically a two-step process. The SO_2-containing flue gas from a boiler is bought into contact with alkali reagent, usually calcium hydroxide. The reaction products are dry and subsequently collected in a particulate control device such as a bag house or an ESP. Consequently, dry waste is produced with handling properties similar to fly ash. The reagent can be delivered to the flue gas in a slurry form (spray-dryer absorption) or as a dry power (dry-injection absorption).

9.4.1 Spray-dryer absorption

Figure 9.6 illustrates the process flow diagram of a spray-dryer absorption process. Hot flue gas meets a mist of finely atomized lime slurry, prepared in a ball-mill slaker, then into a spray dryer. Rotary atomizers or two-fluid nozzles are employed to finely disperse the lime slurry into the flue gas. Typically, the spray dryer is operated at a lime stoichiometry of 0.9 (for low S-content coal) to 1.5 (for high S-content coal). The amount of water fed into the dryer is carefully controlled to avoid complete saturation of the flue gas. Simultaneous heat and mass transfer, between SO_2 in the flue gas and lime in the slurry, result in the absorption reactions (Eqs. 9.4 and 9.5) and the drying of the gas. Some of the solids will settle to the bottom of the dryer, while the remaining solids will travel to the downstream particulate removal device. Part of the collected solids can be recycled. Typical residence time in a spray dryer is 8 to 12 seconds (EPA, 2002; EPA, 2012).

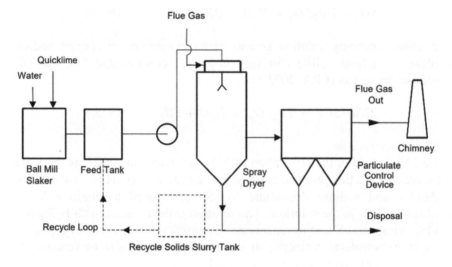

**Figure 9.6 - Schematic of spray-dryer absorption
(EPA, 2002)**

9.4.2 Dry-injection absorption
Dry-injection absorption uses superfine alkali powders, < 325 mesh (< 44μ), to have a large specific area for effective absorption. The main disadvantage is that the typical calcium hydroxide feed rate is three to four times the stoichiometric amount, much higher than that of the spray dryer. Its main advantage is a smaller capital cost (e.g., a dedicated absorber may not be needed).

Figure 9.7 shows a duct-sorbent-injection process in which the alkali reagent, usually hydrated lime, is injected directly into the duct downstream of the boiler's preheater. Water may be injected separately upstream or downstream of the reagent injection point to humidify the flue gas to facilitate the absorption. The distance between these two injection points is critical for effective contacts between the water droplets and the reagent powders. Reagent in a slurry form can also be used for injection instead (EPA, 2002).

298

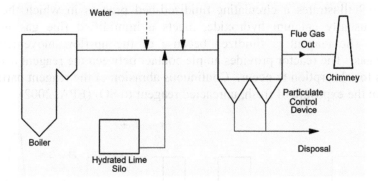

Figure 9.7 - Schematic of a duct-sorbent-injection process (EPA, 2002)

The dry reagent can also be injected directly into the furnace in the optimum temperature region above the flame (Figure 9.8). Due to the high temperature (~1,000 °C), calcium hydroxide will decompose to yield a large specific surface area for absorption. The basic reactions are (EPA, 2002):

$$Ca(OH)_2 \xrightarrow{heat} CaO + H_2O \qquad (9.21)$$

$$CaO + SO_2 + \frac{1}{2}O_2 \rightarrow CaSO_4 \qquad (9.22)$$

The reaction product ($CaSO_4$) and any remaining unreacted reagent will leave the furnace with the flue gas and then enters the particulate removal device.

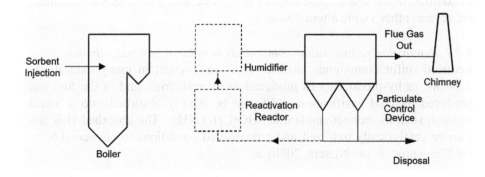

Figure 9.8 - Schematic of a furnace sorbent injection process (EPA, 2002)

Figure 9.9 illustrates a circulating-fluidized-bed process in which the dry reagent, usually calcium hydroxide, meets a humidified flue gas in the reactor. The reagent is fluidized because of the up-flow movement of the flue gas. The reactor provides ample contact between the reagent and the flue gas for absorption to occur. Continuous abrasion of the reagent particles results in the exposure of fresh, unreacted reagent to SO_2 (EPA, 2002).

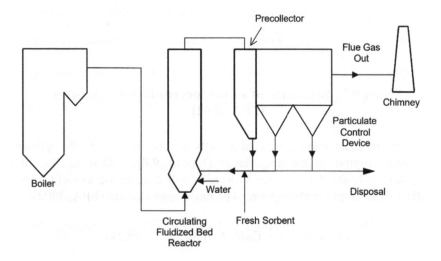

Figure 9.9 - Schematic of a circulating-fluidized-bed process (EPA, 2002)

9.5 Alternatives to End-of-the-Pipe Control

In addition to the end-of-pipe control to remove SO_2 from the flue gas, there are several other viable alternatives.

9.5.1 Removal of reduce sulfur compounds in natural gas and petroleum
Reduced sulfur compounds, such as H_2S, are present in many natural gas deposits, as by-product gases produced in oil refineries, and in the fuel gas produced by coal gasification. H_2S can be readily absorbed into a liquid solution such as monoethanolamine, $HO(CH_2)_2NH_2$. The absorbed H_2S can then be catalytically oxidized under controlled conditions to elemental S by the Claus process (de Nevers, 2000) as:

$$H_2S + \frac{1}{2}O_2 \rightarrow S + H_2O \qquad (9.23)$$

9.5.2 Coal liquefaction and gasification

Coal gasification is to produce a combustible gas from coal through three basic steps: pretreatment, gasification, and gas cleaning. The pretreatment step involves coal pulverizing and washing. Some of sulfur compounds, such as pyrite, can be removed. The pulverized coal is gasified in a pyrolytic reactor. During gasification, the sulfur content in the coal will be converted to H_2S, which is then converted to elemental S by the Claus process (Eq. 9.23).

Coal liquefaction is to change solid coal into synthetic oil. One approach is to gasify the coal (similar to coal gasification), followed by condensation to convert the gas into oil. H_2S can be removed from the synthetic gas. The other approach is to use a solvent to liquefy pulverized coal and then process the liquid into heavy fuel oil (EPA, 2003). Both coal liquefaction and gasification turn coal into a product that can be transmitted through pipelines.

9.5.3 Coal cleaning

Physical coal cleaning has been used for decades to reduce the sulfur and ash content in coal. It utilizes the density differences between the coal and the sulfur bearing impurities. The coal is crushed, washed, and then separated by cyclones, air classifiers, or magnetic separators. Approximately 40 to 90% of the pyrite sulfur content can be removed (EPA, 2003). Chemical coal cleaning is to remove organic sulfur compounds in coal. Several approaches are being developed, including microwave desulfurization and hydrothermal desulfurization (EPA, 2012).

9.5.4 Low-sulfur fuel

Low sulfur coal usually contains 0.5 to 1% S, while high sulfur coal contains 3 to 5% sulfur. Using low sulfur coal can reduce SO_2 emissions. However, both heating value and sulfur content should be taken into consideration, because many low sulfur-content coals have much lower heating values. The other alternative is to substitute a portion of the heat demand supplied by a high sulfur-content coal with natural gas or No. 2 fuel oil which has lower sulfur contents.

9.5.5 Recovery for beneficial reuses

The SO_2 concentrations in the off-gases from metal smelting generally range from 2 to 40%. Using FGD to removal SO_2 for such high concentrations would be costly. Such gases can be economically converted to valuable H_2SO_4 through the following processes (de Nevers, 2000):

$$SO_2 + \frac{1}{2}O_2 \xrightarrow{vanadium\ catalyst} SO_3 \quad (9.24)$$

$$SO_3 + H_2O \rightarrow H_2SO_4 \quad (9.25)$$

9.6 Process Selection and Design Considerations

9.6.1 Process selection

The only sulfur source for SO_x formation from fuel combustion is the fuel itself. Therefore, using fuel of low sulfur content is a direct way to reduce SO_x emissions. However, some fuels with lower S-content may have less heating value. Consequently, emission reductions by using low S-content fuel may not be significant. The first consideration in process selection is if the SO_x concentrations high enough to warrant beneficial reuses of the reaction products. This may only be applicable to waste gas streams of very high SO_x concentrations from metal ore smelting. The fundamental principle of the end-of-the-pipe control is to add alkali agents to react with acidic sulfur oxides. Since the processes use lots of chemicals. The trade-off between the capital cost and the O&M costs should be considered. There are a variety of wet and dry FGD technologies commercially available. Availability and cost of the scrubbing agent should always be taken into consideration. For a specific application, a pilot-scale testing is recommended to evaluate the feasibility of the process and to gather data for site-specific design.

9.6.2 Design calculations

The amount of SO_x formation from fuel combustion is related to the sulfur content of the fuel used. The maximum mass emission and concentration can be estimated from the fuel sulfur content, as in the following example.

Example 9.1 SO_x emissions from coal combustion

Refer to the coal firing boiler in Example 8.1, determine the mass emission rate and the concentration of SO_2 in the flue gas. The coal firing rate is 100 metric ton/hr using the stoichiometric amount of air. Results of coal ultimate analysis (by wt.) are shown below:

7. C: 65.0%
8. H: 8.0%
9. O: 7.0%
10. S: 1.0%
11. N: 1.5%
12. Moisture: 4.0%
13. Ash content: 13.5%

Solution:

(a) Mass loading rate of S to the boiler = $(100)(1\%) = 1$ ton/hr = 1,000 kg/hr

(b) Molar flow rate of S to the boiler = $(1,000/32) = 31$ kg-mole S/hr

= the molar flow rate of SO_2 in the flue gas

= 31 kg-mole SO_2/hr = $(31)(32 + 16\times2) =$ 2,000 kg SO_2/hr = 2 ton/hr

(c) From Example 8-2, total moles of the flue gas = 37,510 kg-mole/hr

(d) SO_2 concentration in the flue gas = $(31)/(37,520)$

= 8.26×10^{-4} = 826 ppm

Discussion:

1. This is a worst-case scenario, assuming all the sulfur in the fuel is converted into SO_2.

2. Every one kg of S in the fuel can form 2 kg of SO_2.

Example 9.2 Switching to low sulfur-content coal

A power generation plant is currently using high sulfur-content (1.5% by wt.) coal, having a heating value of 12,000 Btu/lb (5,160 kJ/kg). There is a lower sulfur content (1.0% by wt.) coal available, but having a lower heating value of 9,000 Btu/lb (3,870 kJ/kg). How much SO_2 emission (in percentage), if the plant switches from the existing coal to the new source?

Solution:

Basis: 1 metric ton of the current coal

(a) Mass of S in this 1-ton coal = $(1,000)(1.5\%) = 15$ kg

(b) Heating value of this 1-ton coal = $(5,160)(1,000) = 5.16 \times 10^6$ kJ

(c) Mass of coal from the new source to have the same heating value

= $(5.16 \times 10^6$ kJ$) \div 3,870$ kJ/kg = 1,333 kg

(d) Mass of S in this new batch of coal = $(1,333)(1\%) = 13.33$ kg

(e) Percentage reduction in SO_2 emission = $(15 - 13.33)/15 = 11.1\%$

Discussion:

1. Mass of S was used in the calculations. The result will be the same if mass of SO_2 was used.

2. The reduction is only 11.1% even the sulfur content of the new coal is one-third less.

9.7 Summary

1. Stationary fuel combustion is the major anthropogenic source of SO_x emissions.

2. The sulfur in SO_x generated from fuel combustion is from the fuel.

3. Using fuel of low sulfur-content is a direct way to reduce SO_x emissions from fuel combustion.

4. SO_x concentrations in flue gases may not warrant resource recovery because the concentrations are typically low.

5. Flue gas desulfurization is essentially a scrubbing/absorption process in which various caustic reagents are used to react with acidic sulfur oxides for their removal.

Bibliography

de Nevers, N (2000). *Air Pollution Control Engineering (2nd edition)*, McGraw-Hill Companies, Inc.

USEPA (1981). *APTI SI 422: Air Pollution Control Orientation Course – Unit 3 Air Pollution Meteorology (3rd edition)*, EPA 450/2-81-017C, prepared by Northrop Services, Inc. for United States Environmental Protection Agency, Research Triangle Park, NC 27711.

USEPA (1992). *Control of Air Emissions from Superfund Sites.* EPA/625/R-92/012, Office of Research and Development, United States Environmental Protection Agency, Washington, DC 20460.

USEPA (1999). *APTI 444: Air Pollution Field Enforcement - Student Manual*, Office of Air Quality Planning and Standards, United States Environmental Protection Agency, Research Triangle Park, NC 27711.

USEPA (2000). *Controlling SO₂ Emissions: A Review of Technologies*, EPA/600/R-00/093, prepared by Srivastava, R.K. for Office of Research and Development, Washington, D.C. 20460.

USEPA (2002). *EPA Air Pollution Control Cost Manual (6th edition),* EPA/452/B-02-001, Office of Air Quality Planning and Standards, United States Environmental Protection Agency, Research Triangle Park, NC 27711.

USEPA (2003). *APTI 452: Principles and Practices of Air Pollution Control - Student Manual (3rd edition),* Air Pollution Training Institute, United States Environmental Protection Agency, Research Triangle Park, NC 27711.

USEPA (2003). *APTI 455: Inspection of Gas Control Devices and Selected Industries - Student Manual (2nd revision),* Air Pollution Training Institute (APTI), Environmental Research Center, United States Environmental Protection Agency, Research Triangle Park, NC 27711.

USEPA (2003). *Flue Gas Desulfurization (FGD) – Wet, Spray Dry, and Dry Scrubbers,* Air Pollution Control Technology Fact Sheet, EPA-452/F-03-034, United States Environmental Protection Agency.

USEPA (2012). *APTI 415: Control of Gaseous Emissions - Student Guide,* Office of Air and Radiation, United States Environmental Protection Agency, Research Triangle Park, NC 27711.

USEPA (2012). *APTI 415: Control of Gaseous Emissions - Student Guide,* Office of Air and Radiation, United States Environmental Protection Agency, Research Triangle Park, NC 27711.

USEPA (2012). *APTI 427: Combustion Source Evaluation - Student Manual (3rd Edition),* prepared by ICES Ltd. for Air Pollution Training Institute, United States Environmental Protection Agency, Research Triangle Park, NC 27711.

Exercise Questions

1. A facility operates several boilers which burn fuel oil. The amount of fuel oil burned at the facility is 150,000 gallons/yr. The fuel oil (specific gravity = 0.90) contains 0.5% of S (by weight). Assuming all the S contained in the fuel oil would be converted to SO_2, and the emission control devices would remove 95% of SO_2 generated during combustion, determine the amount of SO_2 emitted to the atmosphere from these industrial boilers annually (in tons/year).

2. A city has one coal-fired power plant with a 90% efficient SO_2 scrubber and one oil-fired power plant without a scrubber for SO_2 emission control. The coal has a heat content of 13,000 BTU/lb and contains 3.5% (by wt.) sulfur; while the oil has a heat content of 6,000,000 BTU/barrel (one barrel = 42 gallons) and contains 0.9% (by wt.) sulfur. The oil has a specific gravity of 0.92. Assuming that all sulfur is oxidized to SO_2 in the combustion process, determine the SO_2 emission rate per unit amount of electricity produced from each plant (in lb SO_2/BTU).

3. For a coal firing rate of 120 metric ton/hr, what is the minimum amount of air required (in standard m^3/hr) to complete the combustion? Results of coal ultimate analysis (by wt.) are shown below:
   - C: 60.0%
   - H: 7.0%
   - O: 8.0%
   - S: 2.5%
   - N: 1.0%
   - Moisture: 5.5%
   - Ash content: 16.0%

4. For the coal combustion described in Question #3, what would be the flow rate of the flue gas (in kg-mole/hr and standard m^3/hr)?

5. For the coal combustion described in Question #3, what would be the concentrations of CO_2 and SO_2 in the flue gas? What would be the mass emission rate of CO_2 (in ton/hr)?

6. For the coal combustion described in Question #3, if 50% excess air is applied, what would be the flow rate of the flue gas (in kg-mole/hr and stand m^3/hr)?

7. For the coal combustion described in Question #3, if 50% excess air is applied, what would be the concentrations of CO_2 and SO_2 in the flue gas? What would be the mass emission rate of CO_2 (in ton/hr).

Chapter 10

Control of Mobile Source Pollution

Emissions from an individual automobile are low, relative to a smoke stack. However, emissions from numerous automobiles across the country, especially in urban areas, become a big polluter as a whole. Driving a personal vehicle is probably the most "polluting" daily activity of a typical citizen (EPA, 1994).

This chapter starts with an overview of mobile pollution which presents categories of mobile sources, types, amounts, and annual trends of emissions (Section 10.1). Section 10.2 describes emissions from engines and factors affecting their emissions. Section 10.3 discusses the alternatives for control of mobile source emissions.

10.1 Overview of Mobile Source Pollution

10.1.1 Mobile sources of air pollution
Mobile sources of air pollution are divided into two categories by EPA:

- On-road vehicles
  - Passenger cars and trucks
  - Commercial trucks and buses
  - Motorcycles
- Non-road vehicles and engines
  - Aircraft
  - Locomotives
  - Marine vessels
  - Recreation vehicles (snowmobiles, all-terrain vehicles, etc.)
  - Heavy equipment
  - Small engines and tools (lawnmowers, etc.)

10.1.2 Pollutants emitted by mobile sources
The main cause of air pollutant emissions from mobile sources is fuel combustion. Consequently, the main compounds of concern (COCs) are CO, CO_2, hydrocarbons, NO_x, and SO_2. Emissions of particulate matter are also of concern, especially from diesel engines and some off-road vehicles and

equipment using diesel fuel. Table 10.1 tabulates the emissions (in thousand tons) of CO, NO_x, VOCs, and SO_2 in the U.S. in 2017. The contributions of the transportation sector are highlighted; and they are responsible for 54, 59, 21, and 3% of the total emissions of CO, NO_x, VOCs, and SO_2, respectively. Emission of SO_2 from the transportation sector is relatively small, because the sulfur contents in most of the fuels used for mobile sources are low. Reductions of emissions from transportation would help to reduce the corresponding total emissions in CO, NO_x and VOCs. It is interesting to note that the "off-highway" emissions are equally as important as the "highway" emissions. The SO_2 emission from the "off-highway sources" is even larger, which may be attributable to the fact that fuels used in off-highway devices may have high sulfur content. EPA has established a long list of compounds emitted by mobile sources (EPA, 2006).

The percentages of emissions from mobile sources are even greater in urban areas because of more frequent uses of automobiles. The emissions of NO_x and VOCs contribute to the formation of photochemical smog which is a prevailing problem in many metropolitan areas.

Table 10.1 - Emissions (in 1,000 tons) of CO, NO_x, VOCs and SO_2 in the U.S. in 2017

	CO	NO_x	VOCs	SO_2
Stationary fuel combustion	4,065	2,839	519	2,035
Industrial and other processes	4,001	1,282	7,557	534
Transportation - Highway	*18,893*	*3,695*	*1,801*	*27*
Transportation - Off-Highway	*13,269*	*2,660*	*1,656*	*69*
Miscellaneous	19,882	301	4,699	150
Total	60,109	10,776	16,232	2,815
Transportation/Total	**54%**	**59%**	**21%**	**3%**

We have discussed the trends and sectors' contributions to emissions of VOCs, NO_x and SO_2 in previous chapters. Here is a good place to discuss those of CO (CO_2 will be discussed in Chapter 12).

Table 10.2 provides a breakdown of CO emissions from different sectors in the U.S. in 2017. The total CO emissions were approximate 60 million tons in 2017. As shown, fossil fuel combustion (including transportation) is the main source for all these pollutants, except for VOCs. The percentage emissions are 53.5, 33.1, 6.8, and 6.7% from the transportation, miscellaneous, stationary combustion, and industrial and other processes, respectively. Wildfires contributed more than half of the CO emission from the miscellaneous sector. For the transportation sector, high-way vehicles contributed about 60%. Figure 10.1 illustrates the percentages of CO emissions from different sectors in the U.S. in 2017. As expected, emission from the highway vehicles is 31% of the total and that of the off-highway devices is 22%.

Table 10.2 - CO emissions from different source categories in the U.S. in 2017

Source Category	(1,000 tons)	(%)
Stationary fuel combustion	**4,065**	**6.8**
Electric utility	731	1.2
Industral	926	1.5
Others	2,408	4.0
Industrial and other processes	**4,001**	**6.7**
Chemical & allied product manufacturing	129	0.2
Metals processing	610	1.0
Petroleum & related industries	702	1.2
Other industrial processes	584	1.0
Solvent utilization	2	0.0
Storage & transport	8	0.0
Waste disposal & recycling	1,967	3.3
Transportation	**32,162**	**53.5**
Highway vehicles	18,893	31.4
Off-Highway	13,269	22.1
Miscellaneous	**19,882**	**33.1**
Wildfires	10,487	17.4
Others	9,395	15.6
Total	**60,109**	**100.0**

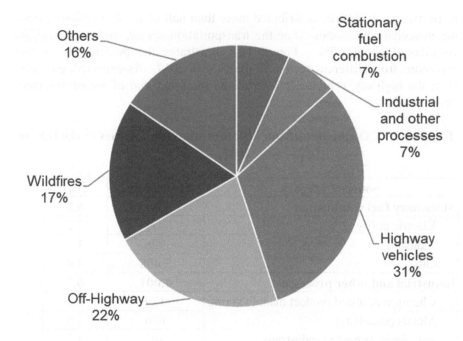

Figure 10.1 - Percentages of CO emissions from different sectors

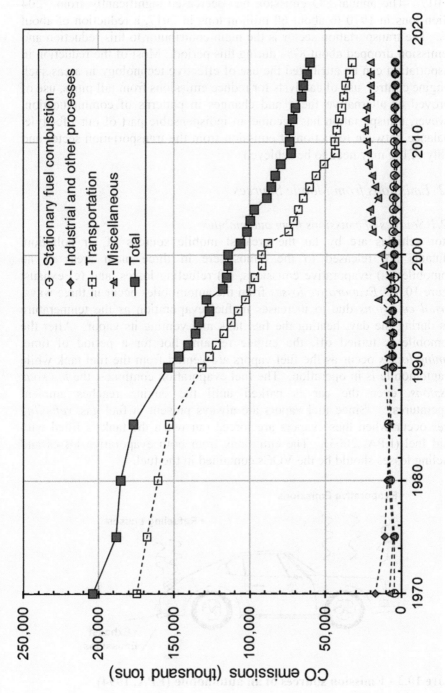

Figure 10.2 - Trends of annual CO emissions in the U.S. (1970 – 2017)

Figure 10.2 illustrates trends of annual CO emissions in the U.S. from 1970 to 2017. The annual CO emission has decreased significantly from >204 million tons in 1970 to about 60 million tons in 2017, a reduction of about 70%. The transportation sector is the main contributor to this reduction and its emission dropped about 87% during this period. Most of the reduction in transportation can be attributed the use of effective technology in areas such as engine control, use of catalysts for reduce emissions from tail pipes, use of improved or alternative fuels, and changes in patterns of communication. However, transportation has become an indispensable part of our life-style. A balance between reduction of emission from the transportation sector and quality of life may need to be achieved.

10.2 Emissions from Mobile Sources

10.2.1 Sources of emissions from automobiles

Motor vehicles are by far the greatest mobile sources of air pollution. Pollutants are released to the atmosphere in three main areas of an automobile: (a) evaporative emissions, (b) refueling losses, and (c) exhaust (Figure 10.3). *Evaporative losses* from the automobiles occur in three ways. *Diurnal emissions* due to increases in fuel evaporation as the temperature rises during the day, heating the fuel tank and venting its vapor. After the automobile is turned off, the engine remains hot for a period of time. *Running losses* occur as the fuel vapors are vented from the fuel tank while the automobile is in operation. The fuel evaporation continues, the *hot soak emissions*, even the car is parked until the engine reaches ambient temperatures. Since fuel vapors are always present in fuel gas, *refueling losses* occur when these vapors are forced out when the tank is filled with liquid fuel (EPA, 2003). The emissions from both evaporative losses and refueling losses should be the VOCs contained in the fuel.

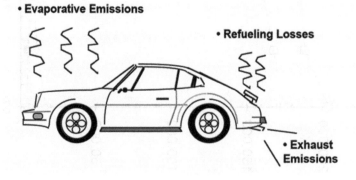

Figure 10.3 - Emission sources of an automobile (EPA, 1994)

The most significant emission from an automobile is resulted from the fuel combustion in its engine that is vented through the exhaust pipe. This exhaust accounts for approximately 90% of the total emissions by an automobile. In addition to unburned hydrocarbons, the exhaust also contains CO, CO_2, NO_x, and some additional hazardous compounds (e.g., acetaldehyde, formaldehyde and 1,3-butadiene) that are not present in the fuel (EPA, 2003). The extent of pollutant emissions depends on the design and operations of the engines.

10.2.2 Reciprocating engines
Reciprocating engines are used in cars or trucks. Combustion in reciprocating engines is not under steady-state. A vehicle can have a 4-, 6-, or 8-cylinder engine. Typically, each cylinder fires once for each two revolutions of the crankshaft. Figure 10.4 shows the typical sequence of the events (intake → compression → power → exhaust). Firstly, the intake valve is open, the piston draws in an air-fuel mixture (intake). Secondly, the intake valve closes and the mixture is compressed on the up stroke (compression). Ignition occurs when the piston is near the top. Thirdly, the power stroke follows as combustion drives the piston down (power). Lastly, the exhaust gases are pushed out on the final up-stroke (exhaust).

Fuel can get into the cylinder of a reciprocating engine in two ways: spark ignition and compression ignition. In *spark ignition* (SI), the air and fuel are premixed before drawn into the cylinder on the down-stroke and the ignited with an electric spark. It is a premixed flame. Diesel engines employ *compression ignition*, in which air is drawn into the cylinder and the fuel oil is injected as the piston approaches the top. The fuel auto-ignites as it is sprayed into the chamber due to the high compression. It is a diffusion flame.

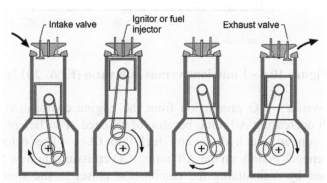

Figure 10.4 - Operating sequence of a reciprocating engine (EPA, 2012)

10.2.3 Factors affecting the pollutant formation and emissions

Figure 10.5 illustrates the hypothetical relationship between the A/F ratio and emissions from an engine. The stoichiometric A/F ratio in that figure is around 16 (mass/mass). The rich-burn refers to the condition in which there is not enough air to burn up all the fuel present. On the other hand, the lean-burn refers to the air present exceeds the stoichiometric amount. An IC engine can operate at one of the three modes: rich-burn, stoichiometric, and lean at different times. As shown in Figure 10.5, when an IC engine operates in the lean mode, the emissions of CO, NO_x, and non-methane hydrocarbon (NMHC) are lower. The fuel economy is also better. However, the power will be relatively poorer. When a larger power is required (e.g., starting, passing, and idling), the IC engine would operate in the rich-mode. In this mode, the fuel economy is worse and the air pollutant emissions are greater.

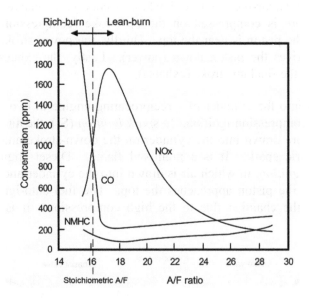

Figure 10.5 - Emissions versus A/F ratio (EPA, 2012)

Carbon monoxide. CO production from the engine combustion increases sharply with decreasing A/F ratio because of absence of sufficient oxygen to completely convert the hydrocarbon fuel into CO_2. The extent of CO formation depends mainly on the A/F ratio. CO emissions in new SI engines are controlled by maintaining the combustion closer to the stoichiometric A/F ratio with the use of a catalyst control system (EPA, 2003).

314

Nitrogen oxides. As mentioned in Chapter 8, formation of NO_x depends on the peak combustion temperature, the residence time of the air-fuel mixture flame, and the level of oxygen present in the combustion zone. For SI engines, combustion efficiency increases with higher temperatures, which would enhance NO_x formation. However, NO_x formation can be reduced by operating in a lean-burn mode, retarding or delaying the spark, or lowering the engine compression ratio. Delaying the spark is to let ignition occur late in the compression stroke so that the time available for production of higher temperatures will be reduced. In the second event of IC engine operation (compression), the piston moves upward and the air-fuel mixture is compressed. Engine compression ratio is the relationship between the initial and final combustion chamber volumes. Lowering this ratio will produce lower combustion chamber temperatures (EPA, 2003).

Hydrocarbons. Unburned hydrocarbons (HC) are those fuel components escapes the combustion. The escapes can be caused by quenching of the flame near cold cylinder walls, near the thin gaps on the combustion chamber, absorption of fuel components on carbon deposits and lubricating oil, and from liquid fuel (particularly during the cold start). The extent of HC emission depends mainly on the A/F ratio, cylinder compression ratio, engine speed, and spark timing. Approximately 1.5 to 2% of the gasoline fuel leaves an SI engine unburned (EPA, 2003).

Particulate matter. Diesel engines have a larger potential to generate particulate matters (PM). The majority of the particulate matters are unburned carbon particles, followed by soluble organic fraction (SOF) which consists of unburned HC that have condensed into liquid droplets or condensed onto the unburned carbon particles. Burning fuel contains sulfur may generate sulfate particles. Reductions in PM emissions are being achieved with improvements in combustion and fuel systems, introduction of electronic fuel system and increasing fuel injection pressures (EPA, 2003).

10.3 Control of Emissions from Mobile Sources

10.3.1 Control of evaporative and fueling losses
Figure 10.6 illustrates a canister system which is to reduce evaporative emissions from a vehicle. When a car is parked, gas vapors from the engine and fuel tank flow into the canister. When the car is moving, vapors in the canister flow back into the engine to be combusted.

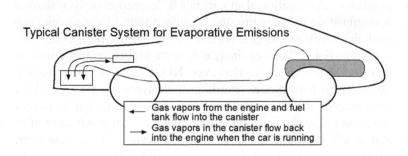

Figure 10.6 - Canister to reduce evaporative losses from a vehicle (EPA, 1994)

Stage II vapor recovery systems in a gas station collect gasoline vapors that are forced out the gas tank during fueling with a vapor recovery nozzle. There are either flexible bellows around the nozzle spout or a port outside it. A vacuum is introduced to collect vapors entering the front end of the nozzle and transport the vapors to an underground gasoline storage tank (Figure 10.7).

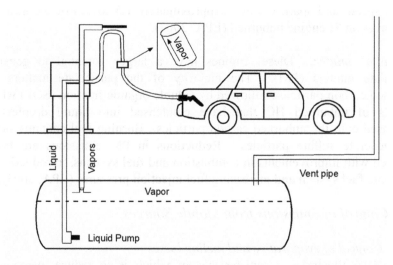

Figure 10.7 - Stage II gasoline vapor recovery system (EPA, 2003)

10.3.2 Control of tail-pipe emissions
Figure 10.8 illustrates a control system to reduce tail-pipe emissions. It generally includes a catalytic converter which hosts "three-way" catalyst, a computer, and an oxygen sensor. The three-way catalyst serves three

functions: (i) convert CO to CO_2, (ii) covert hydrocarbons to CO_2 and H_2O, and (iii) reduce NO_x to N_2 and O_2. The oxygen sensor and computer control are to optimize the efficiency of the catalytic converter.

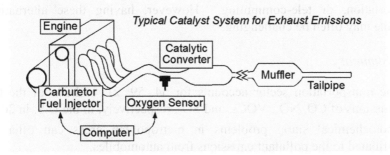

Figure 10.8 - Exhaust control for an automobile (EPA, 2003)

10.3.3 Reformulated gasoline

Reformulated gasoline (RFG) is gasoline blended to burn more cleanly than conventional gasoline. Use of RFG will reduce smog-forming and toxic pollutants in the ambient air. RFG is required in cities with high smog levels and is optional elsewhere. The RFG program was mandated by the 1990 Clean Air Act Amendment.

Reid vapor pressure (RVP) is a common measure of volatility of gasoline. EPA regulates the RVP of gasoline sold at retail stations during the summer ozone season (June 1 to September 15) to reduce evaporative emissions from gasoline. Oxygenates are fuel additives that contain oxygen, usually in the form of alcohol or ether. They can enhance fuel combustion and thereby reduce exhaust emissions, especially CO, and some also boost gasoline octane number. The CAA requires use of oxygenated gasoline in areas where their winter time CO levels exceed NAAQS. In addition, the Mobile Source Air Toxics (MSAT) rules to reduce emissions of HAPs from cars and trucks. The Tier 2 Gasoline Sulfur program, finalized in 2000, was created to reduce sulfur content of gasoline by up to 90%.

10.3.4 Alternative fuels

According to EPA, alternative fuels include gaseous fuel such as hydrogen, natural gas, and propane; alcohols such as ethanol, methanol, and butanol; vegetable and waste-derived oils, and electricity. They may be used in a dedicated system that burns a single fuel, or in a mixed system with other fuels including traditional gasoline or diesel, such as in hybrid-electric or flexible vehicles. Some of these alternative fuels are renewable fuels. Use of alternative fuels may or may not reduce emissions of air pollutants.

317

10.3.5 Other measures

For a given automobile, the emission is basically proportional to the distance it travelled. Reducing the vehicle usage should reduce the pollutant emissions. It can be achieved by carpooling, using more public transportation, or tele-commuting. However, having these alternatives available may often be challenging.

10.4 Summary

1. The transportation sector accounts for 54, 59, 21, and 3% of the total emissions of CO, NO_x, VOCs, and SO_2, respectively, in the U.S. in 2017.

2. Photochemical smog problems in metropolitan areas can often be attributed to the pollutant emissions from automobiles.

3. Many technical measures have been developed to reduce formation and emission of pollutants from mobile sources.

4. There are EPA programs on reformulated gasoline, Reid vapor pressure, winter oxygenates, gasoline sulfur, and mobile source air toxics in place to help emission reductions.

5. Non-technical measures, such as car-pooling and telecommuting, are also viable alternatives to cut mobile source emissions.

Bibliography

de Nevers, N (2000). *Air Pollution Control Engineering (2nd edition)*, McGraw-Hill Companies, Inc.

EPA (1994). *Automobile Emissions: An Overview*, Fact Sheet OMS-5, EPA 400-F-92-007, Office of Mobile Sources, U.S. Environmental Protection Agency.

EPA (2006). *Expanding and Updating the Master List of Compounds Emitted by Mobile Sources - Phase III*, EPA 420-R-06-005, Office of Transportation and Air Quality, U.S. Environmental Protection Agency.

LADCO (2009). *APTI 400: Introduction to Air Toxics - Student Manual*, Lake Michigan Air Directors Consortium (LADCO), Rosemont, Illinois 60018.

USEPA (1999). *Nitrogen Oxides (NOx), Why and How They Are Controlled*, EPA 456/F-99-006R, Office of Air Quality, United States Environmental Protection Agency, Research Triangle Park, NC 27711.

USEPA (2003). *APTI 452: Principles and Practices of Air Pollution Control - Student Manual (3rd edition)*, Air Pollution Training Institute, United States Environmental Protection Agency, Research Triangle Park, NC 27711.

USEPA (2012). *APTI 415: Control of Gaseous Emissions - Student Guide*, Office of Air and Radiation, United States Environmental Protection Agency, Research Triangle Park, NC 27711.

USEPA (2012). *APTI 427: Combustion Source Evaluation - Student Manual (3rd Edition)*, prepared by ICES Ltd. for Air Pollution Training Institute, United States Environmental Protection Agency, Research Triangle Park, NC 27711.

USEPA (2014). *APTI 418: NOx Emission Control from Stationary Sources - Student Manual*, prepared by NESCAUM in 2009 (revised by Doyle B.W. & Slot S. in 2014) for United States Environmental Protection Agency.

Exercise Questions

1. The USEPA estimated that the CO emission of average passenger cars was 20.9 g/mile (13.06 g/km) in 2000. In 2000, USEPA started to implement more stringent emission standards for light-duty vehicles and all the new cars and light trucks were required to meet the CO emission standard of 3.4 g/mile (2.1 g/km) by the end of 2007.

   (a) Compute CO emission in g/mile (g/km) for an automobile traveling at 55 mi/hr (88.5 km/hr) on a highway. The exhaust flow rate is 300 ft^3/min (8.5 m^3/min) and the exhaust contains 0.08% of CO by volume after the catalytic converter (all based on 680 °F (360°C) and 1 atmosphere)

   (b) Will this vehicle meet the new CO emission standard?

2. Consider an urban area with a population of 150,000 and 50,000 vehicles, trafficking within a 100 km^2 (10 km × 10 km) area, with an average driving distance of 12 km during the two-hour period from 8 to 10 a.m. daily. Assuming that each vehicle emits 2.0 g/km of CO and the prevailing wind (4 m/s) blows perpendicular to this square area carrying a negligible concentration of CO, estimate the ambient CO concentration in that area during that time period (the mixing height during that time of the day = 400 m) by using the area-source box model.

319

Chapter 11

Fugitive Emissions and Control

Fugitive VOCs emissions from petroleum refineries and chemical manufacturing facilities can be significant. Sources of these fugitive emissions are mainly leaks from equipment and components of pressurized systems. Control of fugitive emissions can be achieved through equipment practices and work practices. Leak detection and repair (LDAR) program is a good work practice to manage and control fugitive emissions from equipment leaks.

This chapter start with an introduction section which provides definition of fugitive emission, relevant regulations, emission sources and general control approach (Section 11.1). Section 11.2 provides details about the LDAR program which includes its objective, elements and Federal Reference Method 21 for VOC surveying. Section 11.3 describes different types of emission factors and provides a real-life example about how emission factors were derived.

11.1 Introduction

11.1.1 Definition
The EPA defines fugitive emissions as "those emissions which could not reasonably pass through a stack, chimney, vent, or other functionally equivalent opening". The decision to consider an emission source as fugitive or non-fugitive is a factual determination made by the permitting authority.

With regards to air pollution control, control of fugitive emissions has been focused on volatile organic compounds (VOCs) and volatile hazardous air pollutants (VHAPs). Fugitive emissions of GHGs gain increasing attention. For the VOCs and VHAPs, the EPA has determined that leaking equipment are the largest sources of emissions from petroleum refineries and chemical manufacturing facilities. Approximately 70,400 and 9,400 tons/year of VOCs and VHAPs, respectively, have been emitted from equipment leaks. They exceed emissions from storage vessels, transfer operations, or process vents (EPA, 2007). This chapter discuss the major sources of equipment leaks, detection of the leaks, estimate of fugitive emissions, and alternatives to reduce the fugitive emissions from equipment leaks.

11.1.2 Regulations affecting fugitive emissions

Regulations relevant to fugitive emissions from equipment leaks can be found in various EPA regulations, mainly in NSPS, NESHAPs, and hazardous organic NESHAPs (HON). Under these regulations, various industries need to control their fugitive VOC emissions. For example, fugitive emissions from the Synthetic Organic Chemical Manufacturing Industry (SOCMI) accounted for >35% of its total VOC emissions, the HON regulations were implemented to control them (EPA, 2000).

Based on vapor pressure, EPA has classified VOCs into four categories (see Table 11.1). Since the vapor pressure is a function of temperature, the temperature of a process stream will affect the category under which a VOC will belong to. Equipment in VOC service refers to it containing/contacting a process fluid that is at least 10% VOC by wt. If the fluid is in the gaseous or vapor state at operating conditions, the equipment is in gas/vapor service (EPA, 2000).

Table 11.1 - Categories of VOCs (EPA, 2000)

Category	Vapor Pressure
Very volatile	>380 mm-Hg
Volatile	0.1 to 380 mm-Hg
Semi-volatile	0.1 to 10^{-7} mm-Hg
Non-volatile	<10^{-7} mm-Hg

11.1.3 Sources of Fugitive Emissions

A typical refinery or chemical plant can emit 600 to 700 tons of VOCs per year from leaking equipment, mainly because the equipment is pressurized. The following components are the primary sources of emissions subject to leak regulations (EPA, 2007):

- *Pumps:* They are used to transfer fluid. Leaks typically occur at the seal.
- *Valves:* They are used to control flow of fluid. Leaks usually occur at the stem or gland area of the valve, caused by a failure of the valve packing or O-ring.
- *Connectors:* They are used to join piping and process equipment. Common ones include flanges and fitting and gaskets/blinds are usually installed between the flanges to provide the seal. Leaks are commonly caused from gasket failure and improper installation of the flange.

- *Sampling components*: They are used to obtain samples within a process. Leaks commonly occur when the sampling line is purged for sampling.
- *Compressors*: They are used to increase fluid pressure and provide motive force. Leaks most often come from the seals.
- *Pressure relief devices*: They are used to protect equipment from exceeding the maximum allowable working pressure. Common examples are rupture discs and pressure relief valves (PRVs). Common causes for leaks from PRVs include improper valve seating, operating too close to the set point, or worn or damaged seals, while that for rupture discs is improper installations of disc gaskets.
- *Open-ended lines* (OELs): They are pipes or hoses open to the atmosphere. The leaks occur at the point of the line open to the atmosphere.

Previous EPA studies found that valves and connectors accounted for >90% of the emissions from leaking equipment. Newer information suggests that OELs and sampling connections may account for as much as 5 to 10% of the total emissions (EPA, 2007).

11.1.4 Control of Emissions from Equipment Leaks

Since there are different causes for equipment leaks, no single emission control technique is applicable for all types of equipment leaks. The approaches, in general, can be used to control emissions from equipment leaks: (i) equipment practices and (ii) work practices.

Equipment practices involve uses of equipment to eliminate fugitive emissions. Examples include leak-less technology for pumps and valves; caps, plugs, and blinds for OELs; soft seats for PRVs; closed-loop sampling; and enclosure of seal area/vent to a combustion control device (EPA, 1994).

Work practices refer to plans and procedures taken to reduce fugitive emissions. The primary work practice is leak detection and repair (LDAR) of potential emission sources. Many facilities are required, by many NSPS, NESHAPs, SIPs, Resource Conservation and Recovery Act (RCRA), and other state or local requirements to implement LDAR programs. More discussions on LDAR are provided in the next section.

11.2 LDAR Program

11.2.1 Objectives of LADR programs

A well-design and implemented site LADR program would have the following benefits (EPA, 2007):

- Reducing losses of product/material
- Increasing safety for facility personnel
- Decreasing exposure for the neighboring community
- Potentially reducing emission fees
- Avoiding enforcement actions

11.2.2 Elements of an LDAR program

The following of the five elements of an LDAR program (EPA, 2007):

- Identifying components
- Leak definition
- Monitoring components
- Repairing leaking components
- Record keeping

Each regulated component should be assigned a unique identification (ID) number, recorded in a log, and located on the piping and instrumentation diagrams (P & ID) or process flow diagrams. Monitoring intervals vary according to the applicable regulation and the component type. They can be as frequent as weekly. The primary method for monitoring to detect leaking equipment components is Federal Reference Method 21 (FRM 21), which can be found in Appendix A of 40 CFR 60.

A leak is detected whenever the measured concentration by an approved portable analyzer (see section below) exceeds the threshold standard (i.e., leak definition) for the applicable regulations. Leak definitions vary by regulation, component type, service type (e.g., light liquid, heavy liquid, gas/vapor), and monitoring interval. Most NSPS have a liquid definition of 10,000 ppm, while many NESHAP use 500 or 1000 ppm as the leak definition (EPA, 2007; EPA, 2008). A leak can also be defined based on visual inspections or observations.

Leaking components should be repaired as soon as practicable, but not later than a specified number of calendar days after the leak is detected. Detailed records for leak detection and repair should be maintained and updated.

11.2.3 Federal Reference Method 21

FRM 21 specifies a hand-held instrument to be used in determining VOC leaks from process equipment. It defines analyzer performance specifications, but did not recommend specific analyzers or manufacturers. The specifications include monitor response, measurement range, scale resolution, response time, safety, probe dimensions, response factor, and accuracy (EPA, 2017).

11.3 Protocol for Equipment Leak Emission Estimates

There are many equipment components at a specific facility. To quantity fugitive emission from each leaking component is impractical, if not impossible. Using emission factors (EFs), coupled with activity factors, to estimate fugitive emissions from a specific facility or an industry sector as a whole is a commonly-adopted approach. The author of this book conducted a study in 2012 to develop EFs for uses of estimating fugitive emissions from equipment of California natural gas industry. The project was funded by California Energy Commission (CEC). Approaches and findings of the study were extracted and used here to illustrate types, derivation, and uses of EFs for fugitive emissions from equipment leaks. Details can be found in CEC (2012) and Kuo et al. (2014).

11.3.1 Types of emission factors

A number of EFs and correlation equations have been generated for estimation of fugitive emissions from equipment leaks (EPA, 1995; EPA, 1996; API, 2004). These EFs can be categorized into three levels: facility, equipment, and component (EPA, 1995; API, 2009). Using a facility-level EF to estimate methane emission is the simplest approach, requiring only information on type and capacity of a specific facility. The equipment-level EF approach is one level up and requires an accurate count of all major pieces of equipment. The component-level EFs can be further categorized into four groups, in the order of increasing data requirements and increasing accuracy for emission estimates: average EF, screening ranges, correlation-equation, and unit-specific correlation (EPA, 1995; API, 2004).

Use of component-level average EFs for emission estimate needs only the population data of components at a given facility while the other three component-level EFs approaches require screening data from an LDAR program. Screening data are collected by using a portable monitoring instrument to sample air from potential leak interfaces on individual equipment components. A *screening value* (SV) is a measure of the concentration of leaking compounds in the ambient air, typically in parts per million by volume (ppm) by a portable instrument. The instrument typically has a lower limit (the detection limit) and a higher limit (the *pegged SV*). The screening-ranges approach divides the SVs into two or more ranges and an EF is assigned to each range. In older regulations, 10,000 ppm was often used as the leak definition (EPA, 1996). The approach is also known as the "leak/no leak" approach, with the "leakers" having SVs ≥10,000 ppm and "non-leakers" having SVs <10,000 ppm. The correlation-equation approach uses both the SV of a leaking component and the established correlation equations between SVs and leak rates for emission estimate. Each correlation equation is specific to the component type. Using this approach, the emission rate for each leaking component is calculated individually and the facility needs to keep track of the findings from the LDAR activities (API, 2009). The unit-specific correlation approach is similar, except the correlations are unit-specific and/or site-specific. This approach can be expensive and would seldom be justified developing unit-specific correlations to support estimates for GHG emissions (API, 2004).

11.3.2 Derivations of EFs
The portable detection devices used in the 2012 CEC study had a maximum detection limit of 50,000 ppm. The data sets were further divided into four ranges: 0 to 999, 1,000 to 9,999, 10,000 to 49,999, and >50,000 ppm. The first group, "0-999 ppm", was designated as such because all the measured leak rates in this group were below the detection limit of 0.01 ft^3/min (Note: 1 ft^3/min of CH_4 = 59.98 lb/d = 1.316 kg/hr = 9.906 tonne/yr). The second group, "1,000-9,999 ppm" reflects the typical leak definition, 10,000 ppm, in earlier regulations. The third group (10,000 to 49,999 ppm) and the last group (>50,000 ppm) were chosen to take advantage of the maximum detection value (50,000 ppm) of the portable detection devices used in the study.

The screening-ranges EFs are summarized in Table 11.2. As shown, the geometric means of the leak rates increase with SVs.

Table 11.2 - Leak rates (in ft^3/min) versus SVs (Kuo et al., 2014)

SV (ppm)	100 - 999	1,000 - 9,999	10,000 - 49,999	≥50,000
Flanges	0.005	0.0067	0.0293	0.0747
Manual Valves	0.005	0.005	0.0136	0.0197
OELs	-	0.005	0.005	0.2874
Seals	0.005	0.0527	0.1217	0.3907
Threaded Connections	0.005	0.0068	0.0143	0.0168

325

Table 11.3 tabulates the ranges of component leak rates along with the pegged EFs for ≥10,000 and ≥50,000 ppm. The pegged emission rate is the mass emission rate with an SV that is above the maximum detection limit of a portable screening device. As shown, the pegged EFs can be similar or different, based on the detection limits chosen.

Table 11.3 - Pegged EFs in lb CH₄/d (Kuo et al., 2014)

	≥10,000 ppm	≥50,000 ppm
Flanges	2.81	4.48
Manual Valves	1.01	1.18
OELs	7.14	17.24
Others	6.01	6.94
Seals	21.79	23.43
Threaded Connections	0.93	1.01

The component-level correlation-equation approach is often used to estimate an emission rate from a leaking component using a corresponding SV, especially for facilities with LDAR programs in place. The correlation equations are commonly expressed as:

$$LR_{methane} = a \times (SV)^b \quad (11.1)$$

where $LR_{methane}$ = methane leak rate (e.g., in ft³/min)
    a, b = constants developed from the correlation fitting
    SV = screening value (e.g., in ppm)

Table 11.4 tabulated the correlation equations obtained for leak rate in ft³/min for different types of components tested from the 2012 CEC study. As shown, the calculated leak rates for the same SV are different for different component types for 100 ppm ≤ SV ≤ 49,999 ppm.

Table 11.4 - Correlation equations for leak rate in ft³/min (Kuo et al., 2014)

	Correlation Equation	R^2
Flanges	$LR_{methane} = 0.00004 \times (SV)^{0.6549}$	0.4277
Manual Valves	$LR_{methane} = 0.0006 \times (SV)^{0.2811}$	0.1204
Seals	$LR_{methane} = 0.0001 \times (SV)^{0.6816}$	0.4504
Threaded Connections	$LR_{methane} = 0.0006 \times (SV)^{0.3031}$	0.1150

A wellhead is the uppermost part of a natural gas well located at production or storage facilities. For each wellhead surveyed, all components were screened from the ground up. The numbers of wellheads surveyed were 128 and 44 in the production and the storage sectors, respectively. Table 11.5 shows the component-level average EFs of wellheads which were calculated by using the leaking percentages and the mean leak rates from the field survey,

Table 11.5 - Component-level EFs (in tonne CH₄/yr/component) of wellheads (Kuo et al., 2014)

	Production	Storage
Flanges	2.10×10^{-3}	-
Manual Valves	5.85×10^{-4}	9.09×10^{-3}
OELs	-	1.04×10^{-2}
Seals	5.87×10^{-3}	-
Threaded Connections	6.20×10^{-4}	1.27×10^{-3}

To derive a fugitive EF for a wellhead as a whole, the EF of each component was first multiplied by the average number of that component per wellhead. The multiplication products of all components were then summed to become the average EF of a wellhead. The wellhead EFs derived from the 2012 CEC study are tabulated in Table 11.6. As shown, the EFs of the same type of equipment in different sectors may be different.

Table 11.6 - Equipment-level EFs (in tonne CH₄/yr) for wellheads (Kuo et al., 2014)

Production	Storage
3.83×10^{-2}	3.49×10^{-1}

11.3.3 Uses of EFs
Example 11.1 below illustrates uses of EFs to estimate fugitive emission from an equipment.

Example 9.1 Use of emission factors

A gas company has 10 identical natural gas production wells. For each wellhead, there are 10 flanges, eight manual valves, one seal, and 30 thread connections. Use the emission factors in Tables 11.5 and 11.6 to estimate the annual fugitive emissions from these 10 wellheads.

Solution:

(a) Use the component-level EF in Table 11.5 to estimate the fugitive emission from each wellhead

$$= (2.10 \times 10^{-3})(5) + (5.85 \times 10^{-4})(8) + (5.87 \times 10^{-3})(1) + (6.20 \times 10^{-4})(30)$$

$$= 3.97 \times 10^{-2} \text{ tonne } CH_4/\text{yr/wellhead}$$

Fugitive emissions from 10 wellheads $= (3.97 \times 10^{-2})(10)$

$$= 0.397 \text{ tonne } CH_4/\text{yr/wellhead}$$

(b) Use the equipment-level EF in Table 11.6 to estimate the fugitive emission from 10 wellhead

$$= (3.83 \times 10^{-2})(10)$$

$$= 0.383 \text{ tonne } CH_4/\text{yr/wellhead}$$

Discussion: Estimates from two approaches are very similar, because the component counts are very close to the average counts used in deriving the wellhead EF.

11.4 Summary

1. Fugitive VOCs emissions from petroleum refineries and chemical manufacturing facilities can be significant.
2. The sources of these fugitive emissions are mainly leaks from equipment and components of pressurized systems.
3. Control of fugitive emissions can be achieved through equipment practices and work practices.
4. Leak detection and repair program is a good work practice to manage and control fugitive emissions from equipment leaks.
5. Emission factors are commonly used to estimate fugitive emissions from component, equipment, facility, or a sector for emission inventory.

Bibliography

American Gas Association (2008). *Greenhouse Gas Emissions Estimation Methodologies, Procedures, and Guidelines for the Natural Gas Distribution Sector*, a report prepared by Innovative Environmental Solutions, Inc. for American Gas Association.

American Petroleum Institute (2004). *Compendium of Greenhouse Gas Emissions Methodologies for the Oil and Gas Industry*, American Petroleum Institute, February 2004.

American Petroleum Institute (2009). *Compendium of Greenhouse Gas Emissions Methodologies for the Oil and Gas Industry*, American Petroleum Institute, August 2009.

CEC (2012). *Estimation of Methane Emissions from the California Natural Gas System*, CEC-500-2014-072, a report prepared by Kuo, J. for California Energy Commission, Sacramento, CA.

Kuo, J.; Hicks, T.C.; Drake, B.; Chan, T.F. (2015) "Estimation of Methane Emission for California Natural Gas Industry", *J. Air & Waste Management Association* 65(7), 844-55.

LADCO (2009). *APTI 400: Introduction to Air Toxics - Student Manual*, Lake Michigan Air Directors Consortium (LADCO), Rosemont, Illinois 60018.

USEPA (1981). *APTI SI 422: Air Pollution Control Orientation Course – Unit 3 Air Pollution Meteorology (3rd edition)*, EPA 450/2-81-017C, prepared by Northrop Services, Inc. for United States Environmental Protection Agency, Research Triangle Park, NC 27711.

USEPA (1986). *Emission Factors for Equipment Leaks of VOC and HAP*, EPA/450/3-86/002, United States Environmental Protection Agency, Research Triangle Park, NC 27711.

USEPA (1991). *Control Technologies for Hazardous Air Pollutants*, EPA/625/6-91/014, Office of Research and Development, US Environmental Protection Agency, Washington DC 20460.

USEPA (1994). *Control Techniques for Fugitive VOC Emissions from Chemical Process Facilities*. EPA/625/R-93/005, Office of Research and Development, United States Environmental Protection Agency, Washington, DC 20460.

USEPA (1995). *Protocol for Equipment Leak Emission Estimates*, EPA/453/R-95-017, United States Environmental Protection Agency, Research Triangle Park, NC 27711.

USEPA (1999). *APTI 444: Air Pollution Field Enforcement - Student Manual*, Office of Air Quality Planning and Standards, United States Environmental Protection Agency, Research Triangle Park, NC 27711.

USEPA (2000). APTI 380: *Inspection Techniques for Fugitive Emission Sources*, prepared by North Carolina State University for United States Environmental Protection Agency.

USEPA (2000). *APTI SI 380: Introduction to Fugitive Emissions Monitoring*, prepared by North Carolina State University for United States Environmental Protection Agency.

USEPA (2002). *APTI 482: Sources and Control of Volatile Organic Air Pollutants - Student Manual (3rd edition)*, prepared by J.W. Crowder for Air Pollution Training Institute, United States Environmental Protection Agency, Research Triangle Park, NC 27711.

USEPA (2002). *EPA Air Pollution Control Cost Manual (6th edition)*, EPA/452/B-02-001, Office of Air Quality Planning and Standards, United States Environmental Protection Agency, Research Triangle Park, NC 27711.

USEPA (2003). *APTI 452: Principles and Practices of Air Pollution Control - Student Manual (3rd edition)*, Air Pollution Training Institute, United States Environmental Protection Agency, Research Triangle Park, NC 27711.

USEPA (2007). *Leak Detection and Repair - A Best Practices Guide*, Office of Compliance, United States Environmental Protection Agency, Washington DC 20460.

USEPA (2008). Leak Detection and Repair - A Best Practices Guide, Office of Compliance, United States Environmental Protection Agency, Washington DC 20460.

USEPA (2008). *APTI 435: Atmospheric Sampling Course – Student Manual (5th Edition)*, prepared by Tidewater Operations Center of C^2 Technologies for Air Pollution Training Institute, United States Environmental Protection Agency, Research Triangle Park, NC 27711.

USEPA (2012). *APTI 415: Control of Gaseous Emissions - Student Guide*, Office of Air and Radiation, United States Environmental Protection Agency, Research Triangle Park, NC 27711.

USEPA (2013). *APTI 450: Source Sampling for Pollutants - Student Guide*, Office of Air and Radiation, United States Environmental Protection Agency, Research Triangle Park, NC 27711.

Exercise Questions

1. Emission factors for particulate matters (PM) from multiple hearth sewage sludge incinerators are 10 and 4 lb/ton sludge without and with cyclone for particulate removal (from EPA AP-42). A site is burning 50 tons of sludge per day and employs a cyclone system to remove PM from the flue gas before discharge to the atmosphere.

   (a) What is the daily PM emission rate using the EPA's EF value (in lb/day)?

   (b) Based on these two EF values, what is the assumed PM removal by cyclone (in %)?

Chapter 12

Ozone, Acid Rains, and Greenhouse Gases

Many air pollutants can cause regional problems, go across a national border, and even become global issues. Acid rain problems, ozone depletion in the stratosphere, and climate change due to greenhouse gas emissions are good examples

This chapter starts with a discussion on basic photochemical cycle and role of VOCs in the cycle (Section 12.1). Section 12.2 discusses the bad ozone in the troposphere, which is the main component of photochemical smog, as well as the good ozone that is being depleted in the stratosphere. Acid rain is a regional issue because its acidic components can come from air pollutants emitted far away. Section 12.3 covers the basics and effects of acid rain as well as measures to control it. We are also facing a big challenge globally on climate change which greenhouse gases (GHGs) are the main contributors. Basics, emission sources, and control measures of GHGs are covered in Section 12.4.

12.1 Photochemical Smog

Photochemical reactions start with absorption of a photon of energy (hv) by an atom, molecule, ion, or free radical. The equation below illustrates the first step of a photochemical reaction of a species (A):

$$A + hv \rightarrow A^* \qquad (12.1)$$

where A* is the excited state of species A. This excited species may reduce its energy level through dissociation, fluorescence, reaction with another species, or collisional deactivation by transferring energy to another energy absorbing species (EPA, 2002).

The energy for atmospheric photochemical reactions comes from the sun. Light intensity is a critical parameter for photochemical reactions. The wavelength of visible light is from ~390 to 700 nm. Plank's equation relates the energy of a photon (E) with its wavelength (λ) or frequency (v) as:

$$E = hv = h\left(\frac{c}{\lambda}\right) \qquad (12.2)$$

where h = Planck's constant = 6.626×10^{-34} J•s and C = speed of light = 3.00 $\times 10^8$ m/s. As shown, a photon with a shorter wavelength will have a larger frequency and possesses a larger energy.

For the photochemical reactions in the troposphere, the range of light wavelength of interest is from approximately 280 to 730 nm. It is because that (i) ozone in the stratosphere blocks out most of the ultraviolet light shorter than 280 nm and (ii) light with a wavelength greater than 730 nm does not participate in the photochemical reactions of interest. In the U.S., the maximum noon-time light intensity does not vary significantly with latitude during summer months. In the 300 to 400 nm range, the maximum light intensity is ~2×10^{12} photons/m^2/s and remains near this value for 4 to 6 hours. In contrast, that in the winter vary from 0.7×10^{12} to 1.5×10^{12} photons/m^2/s, depending on the latitude, and the duration near this maximum is reduced to 2 to 4 hours. The presence of cloud or aerosols will reduce these maximum values (EPA, 2002).

12.1.1 Basic photochemical cycle

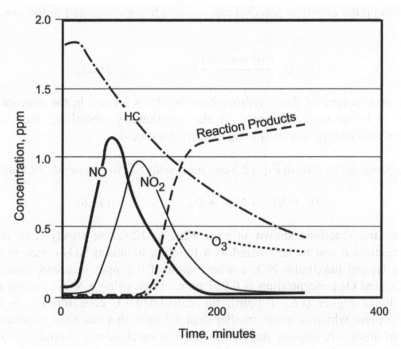

Figure 12.1 - A typical photochemical profile (EPA, 2014)

Figure 12.1 illustrates concentration profiles of NO, NO_2, hydrocarbon (HC), O_3, and reaction products in an ambient air during morning rush hours are a result of automobile emissions. As shown, times of the concentration peaks are in the following increasing order: $NO \rightarrow NO_2 \rightarrow O_3 \rightarrow$ reaction products. Below is a simplified explanation of the chemistry of this smog formation.

Nitrogen oxide (NO) emitted from mobile (and stationary) combustion sources is rapidly converted to NO_2 as:

$$NO + \frac{1}{2}O_2 \rightarrow NO_2 \qquad (12.3)$$

Nitrogen dioxide (NO_2), those newly formed and those directly from the tail pipes, can dissociate by sunlight photolysis to form NO and an atomic oxygen (O) as,

$$NO_2 + hv \rightarrow NO + O \qquad (12.4)$$

The atomic oxygen produced can then react rapidly with oxygen to form ozone with various hydrocarbons working as a *catalyst* (i.e., a substance that speeds up the chemical reaction rate without being consumed in the process) as,

$$O + O_2 \xrightarrow{\quad Hydrocarbons \quad} O_3 \qquad (12.5)$$

The main source of these hydrocarbons is VOCs present in the atmosphere. The hydrocarbons participate in the reaction by absorbing the excess vibrational energy and stabilizing the ozone molecule.

The ozone formed from Eq. 12.5 can react with NO to regenerate NO_2 as:

$$O_3 + NO \rightarrow NO_2 + O_2 \qquad (12.6)$$

Using the reactions shown in Eqs. 12.4 to 12.6, the steady-state ozone concentration can be calculated as a function of initial NO_2 concentration. At a typical maximum NO_2 concentration of 0.1 ppm in urban areas, the calculated O_3 concentration is 0.027 ppm. Even using a NO_2 concentration ten times higher (i.e., 1 ppm), the calculated O_3 concentration is only 0.095 ppm, which is much smaller than 0.5 ppm that has been measured in urban areas. It implies that other reaction mechanisms should also exist to regenerate NO_2 from NO, without consumption of ozone as shown in Eq. 12.6. As will be shown in the subsequent section, that is the role of VOCs plays in the photochemical reactions (EPA, 2002).

Example 12.1   Energy of a photon

Light with a wavelength of 500 nm is green. Determine the frequency and energy of this green light photon.

Solution:

(a) From Eq. 12.2,

$$v = \left(\frac{c}{\lambda}\right) = \left(\frac{3.00 \times 10^8 m/s}{500 \times 10^{-9} m}\right) = 6.0 \times 10^{14} \frac{1}{s}$$

(b) From Eq. 12.2 again,

$$hv = (6.626 \times 10^{-34} J \cdot s)\left(6.0 \times 10^{14} \frac{1}{s}\right) = 3.97 \times 10^{-19} J/photon$$

12.1.2 Roles of VOCs in photochemical smog

Under radiation of light shorter than ~315 nm, ozone can dissociate into an excited oxygen atom (O^*) and an oxygen molecule:

$$O_3 + hv \rightarrow O^* + O_2 \qquad (12.7)$$

Most of the time, the excited O* will release its excess energy and return to its ground state by colliding with N_2 or O_2. However, it occasionally will collide with water molecule to produce two hydroxyl radicals (•OH) as:

$$O^* + 2H_2O \rightarrow 2 \cdot OH \qquad (12.8)$$

Hydroxyl radicals play a key role in atmospheric chemistry. Unlike many radicals formed from organic molecules, they do not react with oxygen. They react with most of other atmospheric compounds, including CO and hydrocarbons (RH), instead. In the notation of RH for hydrocarbons, R represents the alkyl group. The hydroxyl radial will react with hydrocarbons to yield an alkyl radial (R•) as:

$$\cdot OH + RH \rightarrow R \cdot + H_2O \qquad (12.9)$$

The alkyl radical will then react with oxygen to yield peroxyalkyl radicals (RO_2•):

$$R \cdot + O_2 \rightarrow RO_2 \cdot \qquad (12.10)$$

The peroxyalkyl radical can react rapidly with NO to form NO_2 and alkoxy radical (RO•) or yield alkynitrates ($RONO_2$) as:

$$RO_2 \cdot + NO \rightarrow NO_2 + RO \cdot \qquad (12.11)$$

$$RO_2 \cdot + NO \rightarrow RONO_2 \qquad (12.12)$$

One of the alkynitrates potentially formed in the above equation is peroxyacetyl nitrate ($CH_3COOONO_2$), the most common one in the peroxyacyl nitrates (PAN) family, which has a general formula of $RCOOONO_2$. Alkoxy radical can react with oxygen to produce hydroperoxyl radicals (HO_2•) and carbonyl compounds (RCHO):

$$RO \cdot + O_2 \rightarrow HO_2 \cdot + RCHO \qquad (12.13)$$

Finally, the hydroperoxyl radicals can react with NO to regenerate hydroxyl radical and complete the cycle (EPA, 2002):

$$HO_2 \cdot + NO \rightarrow NO_2 + \cdot OH \qquad (12.14)$$

The participation of VOCs in the reactions described in Eq. 12.14 consumes NO and make it less available for ozone destruction as shown in Eq. 12.6. Consequently, this series of reactions cause the build-up of ground-level ozone as well as harmful reaction products.

The hydroxyl radical can also react with NO_2 to form nitric acid as:

$$\cdot OH + NO_2 \rightarrow HNO_3 \qquad (12.15)$$

In summary, this series of reactions include conversion of NO to NO_2, formation of a variety of nitrogen-containing species (e.g., nitric acid and PANs), and accumulation of ozone (EPA, 2002).

12.2 Ozone

Ozone (O_3) molecules occur in the Earth's upper atmosphere and at the ground level. Those present in the stratosphere are considered as good, while those in the ground level are not favorable.

12.2.1 Tropospheric ozone

As discussed earlier, tropospheric ozone is not a primary air pollutant, but is created by photochemical reactions among VOCs, NO_x, and sunlight. It is the main component of the photochemical smog. Ozone is most likely to reach unhealthy levels on hot sunny days in urban environments; but high level ozone concentrations have also been observed in cold months. Ozone can also travel long distances by wind to elevate ozone levels in rural areas.

Ozone is one of the six criteria pollutants under CAA. The ozone pollution control actions are included in vehicle and transportation standards, regional haze and visibility rules and routine reviews of NAAQS in the federal level, SIPs, and other local regulations.

12.2.2 Stratospheric ozone

Stratospheric zone (the "good" ozone") occurs naturally in the upper atmosphere, 6 to 30 miles (10 to 50 km) above the Earth's surface. It forms a protection layer that shields us from harmful ultraviolet (UV) rays from the sun. As a natural process, ozone molecules are constantly formed and destroyed in the stratosphere, the total amount remained relatively constant in the recent past.

However, scientific evidences have shown that the ozone shield has been depleted beyond the natural process since early 1970s. This beneficial ozone is being gradually destroyed by manmade chemicals. At locations where the protective ozone layer has been significantly depleted (e.g., North or South Pole), it is called a "hole" in the ozone layer and the problem is often referred to as *ozone depletion*. This allows more UV radiation to reach the Earth's surface and leads to increased incidences of skin cancer, cataracts, and other health problems. Elevated levels of UV radiation can also cause detrimental ecological effects such as stressing productivity of marine phytoplankton, an essential component of the oceanic food web.

Ozone can be destroyed by radicals such as hydroxyl ($\cdot$OH), nitric oxide (NO$\cdot$), chlorine (Cl$\cdot$) and bromine (Br$\cdot$) radicals. In the stratosphere, the hydroxyl and nitric oxide radicals are naturally occurring, while levels of chlorine and bromine radicals have significantly increased due to human activities. *Ozone-depleting substances* (ODS) are compounds that can travel to the stratosphere, due to its relatively inertness in the troposphere, and release chlorine or bromine radicals by solar UV radiation to destroy ozone. ODS that may release chlorine radicals include chlorofluorocarbons (CFCs), hydrochlorofluorocarbons (HCFCs), carbon tetrachloride (CCl_4), and 1,1,1-trichloroethane ($C_2H_3Cl_3$). Those can release bromine radicals include bromofluorocarbons (halons) and bromomethane (CH_3Br). The discussion below uses trichlorfluromethane ($CFCl_3$, CFC-11) and simplified reactions to illustrate ozone depletion by ODS. A variety of catalytic reaction cycles are actually occurring. $CFCl_3$ can be broken down by solar UV radiation to release chlorine radical as:

$$CFCl_3 + solar\ UV\ radiation \rightarrow Cl\cdot + \cdot CFCl_2 \quad (12.16)$$

The formed chlorine radical can destruct an ozone molecule and form chlorine monoxide radical (ClO•):

$$Cl\cdot + O_3 \rightarrow ClO\cdot + O_2 \quad (12.17)$$

ClO radical can then react with atomic oxygen (O) which formed when solar UV radiation reacts with ozone and oxygen molecules as:

$$ClO\cdot + O \rightarrow Cl\cdot + O_2 \quad (12.18)$$

The net reaction of two reactions shown immediately above is:

$$O_3 + O \rightarrow 2\,O_2 \quad (12.19)$$

Chlorine radical works as a catalyst for ozone destruction because it is reformed each time the reaction cycle is completed; and the end result is an ozone molecule destroyed. In this way, one chlorine atom/radical can participate in many cycles to destroy many ozone molecules before it is removed from the stratosphere. With presence of these substances in the stratosphere, ozone can be destroyed more quickly than it is naturally created.

ODS are regulated as Class I or II controlled substances in the U.S. Class I ODS, including CFCs, halons, 111-TCA, CCl_4, and CH_3Br, have ozone depletion potentials ≥ 0.2 and have been completely phased out in the U.S, with a few exceptions. Class II ODS are all HCFCs and have an ozone depletion potential < 0.2. They are transitional substitutes for many Class I ODS. Most of them will be phased out soon. The most common HCFC currently in use is HCFC-22 (or R-22), a refrigerant in existing air conditioners and refrigeration equipment. Manufacturing of HCFC-22 is responsible for 5% of total fluorinated gases emissions in the U.S. in 2016 (see Section 12.3).

12.3 Acid Rain

Acid rain, or *acid deposition*, includes any form of precipitation with acidic components (i.e., sulfuric or nitric acid) that fall from the atmosphere to the ground in wet or dry form. *Wet deposition* is the sulfuric and nitric acids formed in the atmosphere fall to the ground in the form of rain, snow, fog, or hail. Acidic particles and gases in the atmosphere can also deposit to the surfaces (such as water bodies, vegetation, and buildings) in the absence of moisture as *dry deposition*.

The acidic components in the acid rain come from SO_2 and/or NO_x in the atmosphere. They react with water, oxygen to form sulfuric and nitric acids, then mix with water and other materials before deposit onto the ground. Since wind can blow them over a long distance, acid rain can be a regional, and even an international, problem.

12.3.1 Acid rain basics
Formation of nitric acid (HNO_3) from NO_x, oxygen, and water can be illustrated as:

$$NO_{(g)} + \frac{1}{2}O_{2(g)} \rightarrow NO_{2(g)} \qquad (12.20)$$

$$3NO_{2(g)} + H_2O \rightarrow 2HNO_{3(aq)} + NO_{(g)} \qquad (12.21)$$

Formation of sulfuric acid (H_2SO_4) from SO_2, oxygen, and water can be illustrated as:

$$SO_{2(g)} + \frac{1}{2}O_{2(g)} \rightarrow SO_{3(g)} \qquad (12.22)$$

$$SO_{3(g)} + H_2O \rightarrow H_2SO_{4(aq)} \qquad (12.23)$$

Normal rain is slightly acidic and has pH of about 5.6 because of the dissolution of carbon dioxide into it to form carbonic acid (H_2CO_3) as:

$$CO_{2(g)} + H_2O \leftrightarrow CO_{2(aq)} + H_2O \leftrightarrow H_2CO_3 \qquad (12.24)$$

Acid rain usually has a pH between 4.2 and 4.4, but it can be lower

12.3.2 Effects of acid rain
Both marble and limestone consist primarily of calcium carbonate ($CaCO_3$) and frequently used as construction materials. The reaction between calcium carbonate and the sulfuric acid components in acid rain results in dissolution of $CaCO_3$ into aqueous ions to be washed away:

$$CaCO_{3(s)} + H_2SO_{4(aq)} \rightarrow Ca_{(aq)}^{+2} + SO_{4(aq)}^{-2} + H_2O + CO_2 \quad (12.25)$$

12.3.3 Measures to control acid rain

The component of acid rain responsible for the adverse effects is NO_x or SO_2 in the atmosphere. Only a small portion of SO_2 and NO_x is from natural sources (e.g., volcanos and lightning), most of them from combustion of fossil fuel. The measure to control acid rain is to reduce anthropogenic emissions of NO_x and SO_2, especially from fossil fuel combustion. The Acid Rain Program (ARP), established under Title IV of the 1990 CAAA requires major emission reductions of SO_2 and NO_x from the energy sector.

12.4 Greenhouse Gases

12.4.1 Greenhouse gases

The climate on earth is determined by the balance between the energy received from the sun and the energy emitted back to outer space from the earth and its atmosphere. Gases in the atmosphere that trap some of the outgoing energy, retaining heat in our atmosphere, are called greenhouse gases (GHGs). Water vapor is a GHG. However, the GHGs emitted by human activities are of concern. They are carbon dioxide (CO_2), methane (CH_4), nitrous oxide (N_2O), and fluorinated gases.

The total U.S. GHG emission in 2016 is 6,511 million metric tons of CO_2 equivalent (defined later). CO_2 is the dominant one at 81.5%, followed by CH_4 (10.1%), N_2O (5.7%), and fluorinated gases (2.7%), illustrated in Figure 12.2.

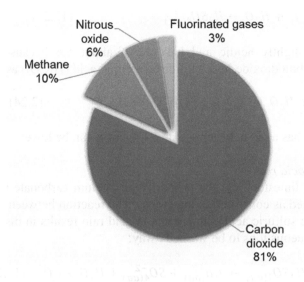

Figure 12.2 - U.S. GHG emissions in 2016 by gas type

12.4.2 Global warming and global warming potential

Global warming refers to the recent and the ongoing rise in global average temperature near the Earth's surface. It has risen by 1.5 °F (0.83 °C) over the past century, and is project to rise further. Now it is a common belief that the global warming is mostly caused by increasing concentrations of GHGs in the atmosphere. Although climate change may be caused by factors other than global warming. Small changes in the average earth's temperature can translate to significant and potentially adverse shifts in weather and climate. Climate change can affect human health, cause changes to forests and other ecosystems, impact our energy supply, and cause severe weather events.

Each of these GHGs can stay in the atmosphere for different durations, ranging from a few years to thousands of years. Due to their long stay, they are well mixed in the atmosphere so that their concentrations are relatively the same all over the world. Different GHGs can have different effects on global warming. Two key factors for the differences are their ability to absorb energy ("radiative efficiency") and their "lifetime" in the atmosphere. By taking these into consideration, *global warming potential* (GWP) was developed to allow comparisons of the global warming impacts of different GHGs. A gas with a higher GWP would absorb more energy per unit mass than those with lower GWPs, and it thus would contribute more to global warming on a per-unit-mass basis. The GWP of CO_2 is unity, which serves as the yard stick.

Table 12.1 tabulates the values of lifetime and GWP of GHGs. The GWP values are based on the most recent assessment report in 2014 (i.e., the Fifth Assessment Report) by the Intergovernmental Panel on Climate Change (IPCC) in 2014. The GWP values in literature may be different because multiple approaches were used to account the influence of future warming on the carbon cycle. In addition, the GWP values are being occasionally updated to reflect the changes in scientific estimates of the energy absorption, lifetime of the GHGs, or atmospheric concentrations of GHGs that result in a change in their energy absorption relative to each other.

The United States primarily uses the 100-yr GWP, which is based on the energy by a gas absorbed over a period of 100 years, as a measure of the relative impact of different GHGs. However, it can be different in some metrics used by the scientific community. Using a 20-yr GWP would prioritize gases with shorter lifetimes than CO_2 in the atmosphere. For examples, the GWP values for CH_4 are 84 - 87 and 28 - 36 for the 20-yr and 100-yr cycles, respectively, because of its shorter life time. On the other hand, the 100-yr GWPs of fluorinated gases would be larger than their 20-yr GWPs.

Table 12.1 - Lifetime and GWP of GHGs

	Lifetime in Atmosphere (yr)	GWP
CO$_2$	varies	1
CH$_4$	12	25
N$_2$O	114	298
Flourinated gases		
PFCs	2,600 - 50,000	7,390 - 12,200
HFCs	up to 270	up to 14,800
NF$_3$	740	17,200
SF$_6$	3,200	22,800

12.4.3 Sources of GHGs

The EPA has established its Inventory of U.S. Greenhouse Gas Emissions and Sinks. It complies with the reporting standards under the United Nations Framework Convention on Climate Change (UNFCCC). It should be noted that the GWP values used are those in IPCC's Fourth Assessment Report in 2007, based on the UNFCCC's guidelines.

Emissions from the electricity, transportation, industry, commercial & residential, and agriculture sectors are 29, 29, 22, 11, and 9% of the total U.S. GHG emissions in 2016, respectively.

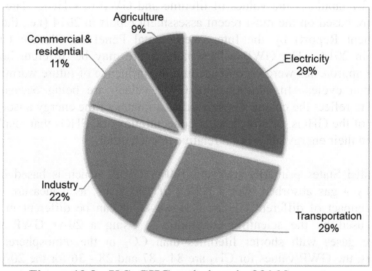

Figure 12.3 - U.S. GHG emissions in 2016 by sectors

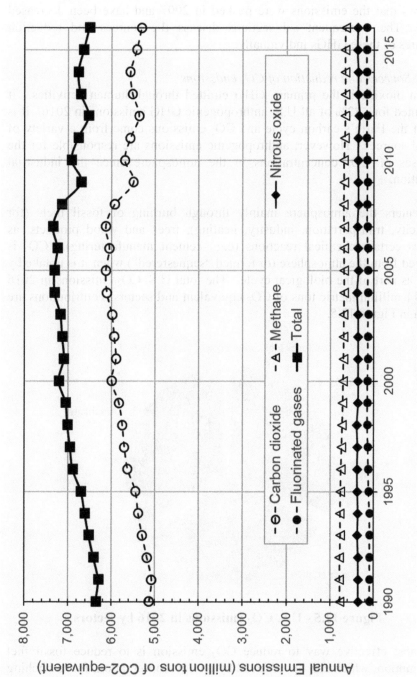

Figure 12.4 - Trends of annual U.S. GHG emissions (1990 to 2016)

Figure 12.4 depicts the annual trends of GHG emissions from 1990 to 2016. It shows that the emissions were peaked in 2007 and have been decreased since. The subsequent sub-sections discuss the sources and reduction measures of these GHGs individually.

12.4.4 Sources and reduction of CO_2 emissions

Carbon dioxide is the primary GHG emitted through human activities. It accounted for ~82% of all U.S. anthropogenic GHG emissions in 2016. It is part of the Earth's carbon cycle and CO_2 emissions come from a variety of natural sources. However, anthropogenic emissions are responsible for the increases of CO_2 concentrations in the atmosphere since the industrial revolution.

CO_2 enters the atmosphere mainly through burning of fossil fuels (for electricity, transportation, industry, heating), trees and wood products, as well as certain chemical reactions (e.g., cement manufacturing). CO_2 is removed from the atmosphere (or termed "sequestered") when it is uptake by plants as part of the biological cycle. The total U.S. CO_2 emissions in 2016 is 6,511 million metric tons of CO_2 equivalent and sectors' contributions are shown in Figure 12.5.

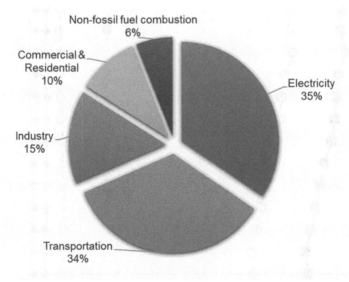

Figure 12.5 - U.S. CO_2 emissions in 2016 by sectors

The most effective way to reduce CO_2 emission is to reduce fossil fuel consumption, which can be achieved by energy conservation, fuel switching and improving energy efficiency. In addition to reduce fossil fuel consumption, carbon dioxide capture and sequestration (CCS) is being utilized to reduce its emission. CCS is a 3-step process that includes (1) capture of CO_2 from power plants or industrial processes, (2) compress the capture CO_2, and (3) then transport through pipelines and inject it into deep underground rock formations for storage and for potential reuses. CO_2 capture is reportedly occurring at >120 facilities in the U.S. and the major beneficial reuse is for enhanced oil recovery (>85% in 2012) with other uses in metal fabrication, and manufacturing of food and beverage as well as pulp and paper.

12.4.5 Sources and reduction of CH_4 emissions
Methane accounted for ~10% of all U.S. anthropogenic GHG emissions in 2016. The total U.S. CH_4 emissions in 2016 is 657 million metric tons of CO_2 equivalent. As shown in Figure 12.6, natural gas and petroleum systems, enteric fermentation, landfills, and manure management are the main sources.

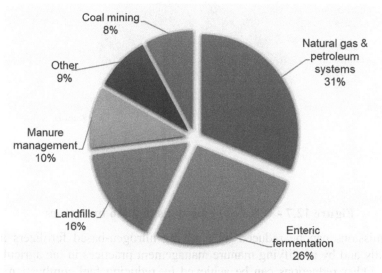

Figure 12.6 - U.S. CH_4 emissions in 2016 by sectors

With regards to emission reduction, employing effective LDAR programs can reduce fugitive emissions from the oil and gas industry. Emission controls that capture landfill gas are a good reduction strategy. Emissions from the agriculture sector can be achieved from better manure management and modifications to animal feeding practices to reduce enteric fermentation. The author of this book conducted a research project for California Air Resources

345

Board (CARB) in 2008, titled *Clearinghouse of Technological Options for Reducing Anthropogenic Non-CO₂ Greenhouse Gases Emissions from All Sectors*, which complied and evaluated technologies for reduction of methane, nitrous oxide, and high GWP gases. The information can be retrieved from http://www.arb.ca.gov/cc/non-co2-clearinghouse/non-co2-clearinghouse.htm and the project report can be downloaded from that website.

12.4.6 Sources and reduction of N₂O emissions
Nitrous oxide accounted for ~6% of all U.S. anthropogenic GHG emissions in 2016. The total U.S. N₂O emissions in 2016 is 370 million metric tons of CO_2 equivalent. As shown in Figure 12.7, agricultural soil management is the main source, accounted for 77% of the total emission.

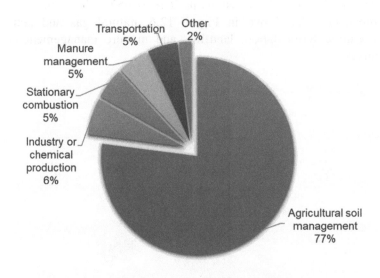

Figure 12.7 - U.S. N₂O emissions in 2016 by sectors

N₂O emissions can be reduced by applying nitrogen-based fertilizers more efficiently and by modifying manure management practices in the agricultural sector. Other reductions can be achieved by reducing fuel combustion, fuel switching and technological upgrade of adipic acid production.

12.4.7 Sources and reduction of fluorinated gases emissions
Different from other GHGs, emissions of fluorinated gases only come from human activities. Their GWPs are much higher, and they are often called high-GWP gases. Their lifetimes are also long, up to thousands of years. There are four main categories of fluorinated gases: hydrofluorocarbons (HFCs), perfluorocarbons (PFCs), sulfur hexafluoride (SF₆), and nitrogen

trifluoride (NF_3). Fluorinated gases accounted for ~3% of all U.S. anthropogenic GHG emissions in 2016. Their total U.S. emissions in 2016 are the equivalent of 173 million metric tons of CO_2. As shown in Figure 12.8, substitution of ozone depleting substances (ODS) is the main source, accounted for 81% of the total emission.

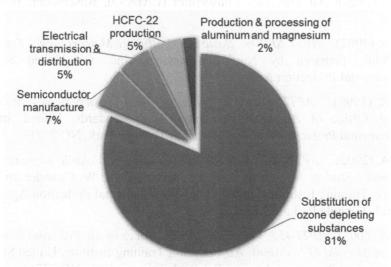

Figure 12.8 - U.S. fluorinated gas emissions in 2016 by sectors

Reductions of fluorinated gas emissions come from better handling the uses of these gases, minimizing leaking of these gases from the systems using them (e.g., air-conditioning, refrigerators, and electricity transmission and distribution systems). Use of substitutes with lower GWPs can also reduce the emissions.

Bibliography

CARB (2008). *Clearinghouse of Technological Options for Reducing Anthropogenic Non-CO_2 Greenhouse Gases Emissions from All Sectors*, prepared by J. Kuo for California Air Resources Board, State of California, Contract No. CARB 05-328 (384 pages) http://www.arb.ca.gov/cc/non-co2-clearinghouse/non-co2-clearinghouse.htm.

Kuo, J. (2012). *Technological Options for Reducing Non-CO_2 Greenhouse Gases Emissions in Handbook on Climate Change Mitigation* (ed. by W.Y. Chen *et al.*), Springer Science, New York, NY.

Kuo, J. (2015). Air Quality Issues Related to Using Biogas from Anaerobic Digestion of Food Waste, California Energy Commission, CEC-500-2015-037, March 2015, 73 pages.

Kuo, J.; Dow, J. (2017). *Biogas Production from anaerobic Digestion of Food Waste and Relevant Air Quality Implication, J. Air & Waste Management Association,* 67(9), 1000-1011. doi: 10.1080/10962247.2017.1316326.

LADCO (2009). *APTI 400: Introduction to Air Toxics - Student Manual,* Lake Michigan Air Directors Consortium (LADCO), Rosemount, Illinois 60018.

USEPA (1982). *APTI SI 409: Basic Air Pollution Meteorology - Student Guidebook,* prepared by Northrop Services, Inc. for United States Environmental Protection Agency.

USEPA (1999). *APTI 444: Air Pollution Field Enforcement - Student Manual,* Office of Air Quality Planning and Standards, United States Environmental Protection Agency, Research Triangle Park, NC 27711.

USEPA (2002). *APTI 482: Sources and Control of Volatile Organic Air Pollutants - Student Manual (3rd edition),* prepared by J.W. Crowder for Air Pollution Training Institute, United States Environmental Protection Agency, Research Triangle Park, NC 27711.

USEPA (2003). *APTI 452: Principles and Practices of Air Pollution Control - Student Manual (3rd edition),* Air Pollution Training Institute, United States Environmental Protection Agency, Research Triangle Park, NC 27711.

USEPA (2012). *APTI 415: Control of Gaseous Emissions - Student Guide,* Office of Air and Radiation, United States Environmental Protection Agency, Research Triangle Park, NC 27711.

USEPA (2012). *APTI 427: Combustion Source Evaluation - Student Manual (3rd Edition),* prepared by ICES Ltd. for Air Pollution Training Institute, United States Environmental Protection Agency, Research Triangle Park, NC 27711.

Exercise Questions

1. Ultraviolet (UV) disinfection is getting popular for water/wastewater disinfection. There are low-pressure UV lamps and medium-pressure UV lamps. The light spectrum of medium pressure lamps is broad, while that of the low pressure lamps is very narrow around 254 nm which is most biocidal. The wavelength of red light is the longest of the visible light. Compare the frequency and energy of UV light with a wavelength of 254 nm with those of red light with a wavelength of 700 nm.

Chapter 13

Other Topics

There are some important air pollution topics that do not fit the flow of the previous twelve chapter well. They are grouped into the last chapter of this book. In my opinion, an environmental professional should be familiar with these topics because they are relevant. Visibility and odor problems are usually not considered health-related (actually some odorants are very toxic). They are often considered as aesthetic issues. However, visibility impairment is mainly caused by particulate matters in the air and many of them originated from anthropogenic sources. In the meantime, presence of odor can be a good indicator that other air pollutants may also be present. We spend more time indoor than outdoor. Although EPA does not regulate indoor air quality, we need to have good indoor air quality in our homes, schools, and offices. Lead is one of the six criteria pollutants, but it seems to be neglected in most of the discussions.

This chapter provides coverages on visibility and haze (Section 13.1), odor (13.2), lead (13.3), indoor air pollution (13.4), and OSHA limits (13.5).

13.1 Visibility and Haze

Visual ranges in the national parks have been substantially reduced by air pollution. During much of the year a veil of white or brown haze hangs in the air blurring the view. Typical visual range in the western U.S. is 60 to 90 miles (100 to 150 km), or about one-half of it would be without haze-causing air pollution. For the rest of the U.S., the typical range is 15 to 30 miles (25 to 30 km), or about one-third of the visual range under natural conditions (EPA, 2006)

Haze is caused when sunlight encounters tiny particles in the air, some of the lights will be absorbed or scattered. The majority of these particles originated from anthropogenic sources. More fine particles will result in more absorption and scattering which reduces the clarity and color of what we see. Some types of particles (e.g., sulfates) will scatter more light, especially during humid conditions.

U.S. EPA and Departments of Interior and Agriculture are working with state, local, and tribal authorities to promote steady improvements in visibility. The federal regulations specifically toward visibility include the

2005 Clean Air Interstate Rule, the 1999 regulations to reduce haze and protect visibility, and the 1992 Acid Rain Rule (EPA, 2006).

13.2 Odor

Odor is a property of a substance that simulates the olfactory organ. Odors from a single compound can be regulated in a manner similar to a criteria pollutant. However, odors generally resulted from a combination of compounds in large numbers. In addition, with respect to air pollution control, an odor is not actually an air pollutant, but rather a property of air pollutants, which can only be detected or measured through its effects on the human organism. A compound having an odor is called an *odorant*. Concentrations of odorants in air are typically low. Even their low concentrations can be measured by using sophisticate instrument, the exact odor strength needs to be determined by a panel of people.

The *odor detection threshold* is the lowest concentration of an odorant that is perceivable by the human sense of smell. The thresholds cannot be accurately detected; they are measured through extensive tests using human subjects instead. The threshold values can be quite different; for example, 100 ppm for methanol and 0.0021 ppm for trimethylamine (C_3H_9N)

The unit to represent the odor strength/concentration is the odor unit. Odor concentration is determined by diluting an odorous air sample with a measured amount of odor-free air and presenting the diluted mixture to an odor panel.

Elimination of odors is the most important part of any odor problem. Air pollutants responsible for an odor should be controlled so that their concentrations in ambient air will not reach their threshold concentrations. Prevention or elimination of sources of odorous compounds should always be the first priority. The destruction or removal of odorants in air is commonly accomplished by direct or catalytic incineration, GAC adsorption, or condensation.

13.3 Lead

Lead is one of the six criteria pollutants. Level of lead in the ambient air has decreased by 99% between 1980 and 2016 as a result of regulatory efforts including the removal of lead from gasoline for motor vehicles. However, sources of lead emissions still exist.

Lead may be emitted into the environment from industrial sources and contaminated sites, such as former lead smelters. Much of exposure comes from human activities (e.g., use of fossil fuels), some types of industrial facilities, and past use of lead-based paint. When lead is released to the air, it may travel long distances before deposit to the ground, where it usually adsorbs onto soil particles.

EPA regulates lead emissions under the NAAQS and several other programs, including NESHAPs and NSR and each state needs to include control of lead in the SIP.

13.4 Indoor Air Pollution

According to EPA, *indoor air quality* (IAQ) refers to air quality within and around buildings and structures, especially as it relates to the health and comfort of building occupants. We usually spend more than half of our time indoors. Therefore, the indoor air quality is as important as, if not more than, the outdoor air quality. Health effects from indoor air pollutants may be immediate or long-term after exposure. With regards to IAQ concerns, the EPA classifies the buildings into three types: (i) home, (ii) schools, and (iii) office and other large buildings. IAQ in homes is more critical to human health, since we stay inside our homes for at least about one third of our lifetime. But the discussion here focuses mainly on IAQ in commercial and institutional buildings.

13.4.1 Sources of indoor air pollutants
The concentration of compounds of concern (COCs) of a building depends on three main factors: (i) input rate from the outdoor air, (ii) interior emission/production rate, and (iii) the air exchange rate. Mass input from outdoor and emission/generation from inside are the total pollutant mass, the actual concentrations will be affected by the frequency of air exchange.

13.4.2 Categories of indoor air pollutants

Although there are numerous indoor air pollutants that may be present in a building, they typically fall into three basic categories: biological, chemical, and non-biological particle.

Biological. Elevated levels of bacteria, viruses, fungi, mold, dust mites, animal dander, and pollen may be present due to inadequate humidity control or water spills/intrusion/flooding/condensation (OSHA, 2011).

Chemical. Chemical pollutants (gases or vapors) can come from the building materials (e.g., walls, floor covering, adhesives, and paints), office equipment (e.g., photocopy machines and printers), maintenance products (e.g., pesticides, cleaning products), human activities (e.g., uses of consumer products, uses of heating devices, and releases from the occupants themselves). Common indoor chemical air pollutants of concern include VOCs, CO, CO_2, formaldehyde, and pesticides. Radon, a radioactive gas, is found in nearly all soils and it can get into any type of buildings. EPA recommends testing all homes below the third floor and schools for radon. In addition, lead paint and asbestos may present in some old buildings.

Non-biological particles. Dust, dirt, or other substances may be carried by the outdoor air into the building. They can also be produced by activities occurred inside the building such as construction and operating of the office equipment (OSHA, 2011).

13.4.3 Effects of indoor air pollution

Good IAQ in a building would contribute to a favorable and productive environment for building occupants. It would give them a sense of comfort, health, and well-being. On the other hand, indoor air pollutants (e.g., VOCs) may have short- and long-term health effects.

The type and severity of health effects resulting from poor IAQ depend on the type and concentration of the pollutant(s) as well as the frequency and duration of the exposure(s). Short-term effects may include allergies, stress, colds, and influenza, while the long-term effects may include respiratory disease, heart disease, and cancer (OSHA, 2011). *Sick building syndrome* (SBS) is used to describe situations in which building occupants experience acute health or comfort-related effects that appear to be linked to time spent in the building, but no specific illness or cause can be identified. In the contrast, *building related illness* (BRI) is used when symptoms of diagnosable illness are identified and can be attributed directly to airborne building contaminants (EPA, 1991).

13.4.4 Measures to reduce exposure indoor air pollution

A 718-page guide, *Indoor Air Quality Guide - Best Practices for Design, Construction and Commissioning* was developed by the American Society of Heating, Refrigerating and Air-Conditioning Engineers (ASHRAE), the American Institutes of Architects, Building Owners and Managers Association International, Sheet Metal and Air Conditioning Contractors' National Association, U.S. Environmental Protection Agency, and U.S. Building Council (ASHRAE, 2009). The project was funded by the EPA and the guide provides practical information covering the full range of IAQ important information useful for architects, engineers, and building owners. The guide can be downloaded free of charge. To have an access of this guide is highly recommended. The guide first describes eight objectives that are needed for improving IAQ and then provides detailed guidance to achieve these objectives. The names of these objectives alone provide a good idea on the control/management measures needed to be taken to improve IAQ. One should consult the guide for details. These eight objectives are:

1. Manage the design and construction process to achieve good IAQ
2. Control moisture in building assemblies
3. Limit entry of outdoor contaminants
4. Control moisture and contaminants related to mechanical systems
5. Limit contaminants from indoor sources
6. Capture and exhaust contaminants from building equipment and activities
7. Reduce contaminant concentrations through ventilation, filtration and air cleaning
8. Apply more advanced ventilation approaches.

Typical engineering controls include the following three measures:

Local exhaust. Using local exhaust (e.g., a canopy hood) to remove point sources of pollutants before they can be dispersed into the indoor air.

General dilution ventilation. A well-designed and functioning heating, ventilating and air-conditioning (HVAC) system would control temperature and relative humidity levels to provide comfort, distribute adequate amounts of outdoor air, and dilute/remove odors and other pollutants.

Air cleaning. Air cleaning primarily involves removal of particulates from the air as the air passes through the HVAC equipment (OSHA, 2011). Other devices such as granular activated carbon should be considered, if needed.

13.4.5 Air exchange rate

Air exchange rate is the rate at which outside air replaces indoor air in a space. It is often expressed as number of air changes per hour (ACH).

13.5 OSHA permissible exposure limits

Occupational Safety and Health Administration (OSHA) is part of the United States Department of Labor. It is to assure safe and healthful working condition for workers by setting and enforcing standards and by providing training, outreach, education and assistance. Many air toxics have their concentration limits in work place.

Permissible exposure limits are established by OSHA and they are legal limits in the U.S. for exposure of an employee to a chemical substance or physical agent such as loud noise. A PEL is usually given as a time-weighted average (TWA). However, some are *short-term exposure limits* (STEL) or *ceiling limits*. A TWA is the average exposure over a specified period, usually a nominal eight hours. This implies that, for limited periods, a worker may be exposed to concentration excursions higher than the PEL, so long as the TWA is not exceeded. A STEL is one that addresses the average exposure over a 15-30 minute period of maximum exposure during a single work shift. A ceiling limit is one that may not be exceeded for any time, and is applied to irritants and other materials that have immediate effects.

Bibliography

ASHRAE (2009). *Indoor Air Quality Guide – Best Practices for Design, Construction and Commissioning.* The guide was developed by American Society of Heating, Refrigerating and Air-Conditioning Engineers (ASHRAE), The American Institutes of Architects, Building Owners and Managers Association International, Sheet Metal and Air Conditioning Contractors' National Association, U.S. Environmental Protection Agency, and U.S. Building Council. The project was funded by U.S. EPA Cooperative Agreement XA-83311201.

EPA (1991). *Indoor Air Facts No. 4 - Sick Building Syndrome (revised),* Office and Air Radiation, U.S. Environmental Protection Agency.

EPA (2006). *How Air Pollution Affects the View,* EPA-456/F-06-001, Office of Air Quality Planning and Standards, U.S. Environmental Protection Agency, Research Triangle Park, NC 27711.

LADCO (2009). *APTI 400: Introduction to Air Toxics - Student Manual*, Lake Michigan Air Directors Consortium (LADCO), Rosemount, Illinois 60018.

OSHA (2011). *Indoor Air Quality in Commercial and Institutional Buildings*, OSHA 3430-04, Occupational Safety and Health Administration (OSHA), U.S. Department of Labor.

USEPA (1995). *APTI 325: Visible Emission Evaluation Procedures Course - Student Manual*, prepared by Environmental Institute for Technology Transfer, University of Texas Austin for United States Environmental Protection Agency.

USEPA (1999). *APTI 444: Air Pollution Field Enforcement - Student Manual*, Office of Air Quality Planning and Standards, United States Environmental Protection Agency, Research Triangle Park, NC 27711.

USEPA (2002). *APTI 482: Sources and Control of Volatile Organic Air Pollutants - Student Manual (3rd edition)*, prepared by J.W. Crowder for Air Pollution Training Institute, United States Environmental Protection Agency, Research Triangle Park, NC 27711.

USEPA (2003). *APTI 452: Principles and Practices of Air Pollution Control - Student Manual (3rd edition)*, Air Pollution Training Institute, United States Environmental Protection Agency, Research Triangle Park, NC 27711.

USEPA (2012). *APTI 415: Control of Gaseous Emissions - Student Guide*, Office of Air and Radiation, United States Environmental Protection Agency, Research Triangle Park, NC 27711.

Index